Victor Horsley, L. C Wooldridge

On the Chemistry of the Blood

and other scientific papers

Victor Horsley, L. C Wooldridge

On the Chemistry of the Blood
and other scientific papers

ISBN/EAN: 9783337387518

Printed in Europe, USA, Canada, Australia, Japan

Cover: Foto ©berggeist007 / pixelio.de

More available books at **www.hansebooks.com**

ON THE

CHEMISTRY OF THE BLOOD

AND OTHER SCIENTIFIC PAPERS

BY THE LATE

L. C. WOOLDRIDGE, M.D., D.Sc.

ASSISTANT PHYSICIAN TO AND CO-LECTURER ON PHYSIOLOGY AT GUY'S HOSPITAL

ARRANGED BY

VICTOR HORSLEY, B.S., F.R.C.S., F.R.S.

AND

ERNEST STARLING, M.D., M.R.C.P.

WITH AN INTRODUCTION BY VICTOR HORSLEY

LONDON

KEGAN PAUL, TRENCH, TRÜBNER & CO. Ltd.

PATERNOSTER HOUSE, CHARING CROSS ROAD

1893

CONTENTS

———◦◦◦———

— — —

PART II. PATHOLOGICAL PAPERS

INTRODUCTION

BY VICTOR HORSLEY

No story in Science is older than the fact that an observer who has the very rare gift of looking at old questions with an impartially scientific and original mind is not only unlikely to receive assistance from those who profess special knowledge of the subject upon which his experimental observations are made, but, on the contrary, will most certainly meet with resistance born of prejudice. The reason of this is not very far to seek, for the continued repetition of a plausible and easy hypothesis, in a few years converts it for many people into an apparent fact, and it then requires an unusually scientific habit of thought to recognise that after all it is but an hypothesis, or at the best but a conceivable mode of explaining certain facts, and that it cannot be allowed to override the contrary weight of evidence derived from a rapidly advancing science. The real difficulty which presents itself is that many who are selected to act as arbiters of the value of any fresh idea have little ability to recognise that all scientific progress lays bare some new fact which, while constituting an addition to our knowledge, or (putting it more correctly) a certain diminution of the previous degree of our ignorance, must necessarily be in opposition to preconceived notions. Thus, when Joule brought forward at the Royal Society his marvellous discovery of the equivalency of heat and work, it was received with such ignorant antagonism that its publication was deferred for years, and Science correspondingly

B

obstructed. So, too, the papers of Waterston, whose brilliant discoveries as an unknown man (anticipating by very many years the equally brilliant demonstrations of Maxwell) were similarly refused publication by the Royal Society. It is impossible to introduce the highly original views of Wooldridge on the Coagulation of the Blood without recognising that, most unfortunately for the progress of physiology, and much to be regretted for other reasons, he, like such pioneers, met with discouraging opposition where, on the contrary, appreciative support ought to have been thankfully accorded. What little help and encouragement he received were given at the earliest stage of his career. Very soon, however, at a far more critical period, his papers were refused publication and returned with comments which, more forcible than polite, revealed the fact that his critics did not understand the foundations even of the questions with which he dealt. Finally, his Croonian Lecture, which, summarising as it does in a remarkably clear manner the bearing of his whole investigations on the Coagulation of the Blood, and explaining most lucidly the relation of his ideas with those universally accepted, would, if it had been given to the world, have raised the author in his lifetime to the position which is now being tardily accorded to him, was actually refused publication. It now lies in the Archives of the Royal Society, and permission has been obtained for it to be copied in order that it may here, for the first time, be included among his published works.

The action of those English scientists upon whom the responsibility of these proceedings rests is the more to be regretted when we remember that Wooldridge's researches were welcomed and respected abroad, and by none more warmly than the great exponent of modern and progressive work, Professor Ludwig. That they have also received from a small number of the Dorpat School the compliment of appropriation without acknowledgment is a strong, though undesirable, proof of the value of his discoveries.

I have felt bound to place the above facts in their true light, firstly, because it is due to the memory of Dr. Wooldridge to show the heavy discouragements and difficulties under which he laboured; secondly, because it may be a means of calling the attention of the Royal Society to an incident in its annals a recurrence of which it ought to be possible to avert in the future.

To turn now to the consideration of Dr. Wooldridge's various scientific communications, it will be found that on page 6 I have grouped them into classes which will, I think, best indicate the sequence of the papers in which his physiological [1] work was published, and will at the same time, I hope, indicate the development of his ideas, and consequently make clearer, if it be necessary, the connection of the latter with the common views which prevail on the subjects he dealt with. At the same time even a brief study of the various articles will show how he gradually evolved his views by a strict adherence, so far as he possibly could, to the great principle which can be so readily detected running all through his work, namely, that of making the ordinary analytical processes of physiological chemistry explanatory, not merely of changes that go on in tissues greatly altered by the action of reagents and repeated physical metamorphoses, such as precipitation and re-precipitation, but, as far as possible, of the truly vital changes. In other words, he applied physiological methods to living tissue, or, where that was impossible, to tissue the degree of the death of which he always most carefully recognised and particularly determined in each of his published papers. In short, he always attempted to employ chemical methods for the elucidation of tissue processes

---

[1] The papers in which he treated questions of Pathology are arranged in a second part.

in the same manner as of recent years Heidenhain has used physical procedures to solve the important bio-chemical changes which are concerned in the formation (*i.e.* 'secretion') of the juices or fluids of the body. Nothing is clearer than that this feeling always actuated Wooldridge, as those who knew him personally are well aware. Even in his very first paper, published in 1881, on 'The Chemistry of the Blood-corpuscles,' this is strikingly evident. Judging in the vague manner that one is necessarily compelled to do, it has always surprised me that this very definite and philosophic principle which thus underlaid his work was not more generally recognised. Much adverse criticism of his ideas, which was founded on an ignorance of this fact, was consequently beside the mark, and particularly so since, as Wooldridge contested, and as recent results have subsequently shown with especial truth, it is most unreasonable to speak of a body which has been dissolved and redissolved many times as a chemical substance of constancy, if its molecule is profoundly unknown. In fact, it is wrong to speak of proteid substances 'isolated' by the physiological chemist as being of known composition and character in exactly the same meaning as is employed in general chemistry when referring to bodies the formulæ of which have been ascertained. While making this prefatory statement it will not be out of place to employ as an illustration a reference to the terminology of this subject, inasmuch as considerable confusion of opinion, and certainly, as many must have seen, a great deal of diversity of thought, have arisen solely because certain terms have been improperly employed in a very restricted sense, and even given to classes of bodies of which not merely are the formulæ unknown, but even their correlation, and indeed coexistence, with other compounds. Perhaps of all terms the one which has been most abused in this particular is the expression 'fibrinogen.' To restrict the term, as some have endeavoured to do, to such bodies as have been described by Schmidt or Hammarsten, is to lose one of the safeguards of scientific language. Wooldridge was most careful never to employ the expression 'fibrinogen' except in its most catholic sense—namely, a substance which is simply a precursor of fibrin; and it surely should hardly be used in any other way, and then only with the qualification of the name of the observer, in an instance where a particular form of such a body may have been isolated by a special method—*e.g.* Schmidt's Fibrinogen, Hammarsten's Fibrinogen, &c. &c.

The investigation of such a complex phenomenon as the formation of an amorphous substance like fibrin out of a living fluid tissue constituted like the blood-plasma, can only be solidly and safely advanced by first determining the earliest precursors of fibrin, next the substances which develop from such precursors, and so on until the complete end-product is arrived at. The general tendency, in the treatment of this subject by physiological chemists, has been to attempt to purify by the ordinary chemical methods *in vitro* what appeared to them to be a precursor of fibrin until something like constancy of composition should be arrived at; but, as Wooldridge urged, any bodies thus ultimately obtained not only could not be considered constant in composition in the sense usually understood by chemists, but also each step in the process of purification necessarily took the observer further and further away from the correlation which must exist between the fibrinogen and the other components of living fluid blood which obviously undergo alteration in the process known as 'clotting.'

Other points of a like nature which may occur in the employment of certain terms will be alluded to when necessary at the moment of use of the particular expression.

I will now classify Wooldridge's physiological communications according to the subjects they elucidate, and according to the various stages of discovery to which they brought him, and following upon this classification it will be my object to put in as succinct a manner as I can what would appear to be the direct relationship existing between his views and those ordinarily received as classical. In a second part the pathological portion of his work will be dealt with.

PART I

IN a special preface by itself I must put first (although it was published in the third of his years of publication) the paper which he wrote dealing with 'The Functions of the Ventricular Nerves of the Mammalian Heart,' because it seems to have been the only research of the kind that he performed, and therefore it does not fall into line with the large series in which the evolution of his ideas of the chemistry of the blood is set forth. It also very fitly takes the first place, being an original research of the greatest importance and opening up a field of investigation absolutely untrodden before. His chemical papers may, I think, be conveniently arranged as follows :—

Class I. The Chemistry of the Corpuscles of the Blood and their Relation to the Process of Coagulation.

Class II. The Discovery and Demonstration of the Occurrence of Coagulation without Ferment Action.

Class III. The true Nature of Fibrinogens.

Class IV. The Croonian Lecture.

Class V. Further Elaboration of his Observations on Coagulation.

Class VI. First Report to the Scientific Committee of the Grocers' Company.

Class VII. Further Papers, including the unfinished Second Report to the Grocers' Company, dealing with the question of the Interaction of Fibrinogens as the Causative Factor in Fibrin Formation.

Prefatory Class

BEGINNING first with the paper which deals with the functions of the Ventricular Nerves of the Mammalian Heart (*i.e.* of the dog), it is important to note that, before Wooldridge took it up, this rather obvious subject of inquiry had never received consideration from the physiological side, although anatomists had for a long time shown the course taken by these nerves. It would, perhaps, be best, considering the thoroughness with which the paper deals with this novel point, to do no more here than to most briefly draw attention to the facts of fresh discovery which it contains. By simply conceived experiments of excitation combined with control experiments of exclusion, Wooldridge showed that the bundles of nerve-fibres which, coming from the cardiac plexuses, ramify in two sets over respectively the front and back surfaces of the ventricles, do not act as the conductors of the inhibitory or acceleratory impulses derived by exciting either the vagus or accelerator nerves. Consequently the first great fact he thus established was that, so far as the spread of these impulses was provided for, it was through the medium of the muscle and other tissue [1] connecting the auricles and ventricles, and not through the channels of the superficial ventricular nerves, and the truth of this he finally established by exclusion experiments. The real function of these same bundles of nerve-fibres, so far as their distribution to the ventricles is concerned, he showed to be *afferent* in character, and experimentally demonstrated that excitation of the central end of any of the ventricular nerves produced, reflexly, slowing of the rate of the heart and rise of the arterial blood-pressure if the nerve stimulated belonged to the set that coursed down the front of the ventricles ; while, on the other hand, he found that excitation of the central end of those on the posterior surface

[1] The harmony between these observations on this question and the experiments of other workers during the last fifteen years is obvious.

frequently evoked acceleration of the heart-rate. Finally, the
nerves which are distributed to the auricle when excited arrest
the auricle, and by transmission through the intervening tissue,
as just stated, cause arrest of the ventricle also.

We may now leave this very interesting and valuable
research and commence the consideration of the papers in which
are embodied his life's work—namely, the Chemistry of the
Coagulation of the Blood, and the deductions which he was
enabled to draw from his observations.

In arranging the classes into which one can place his
numerous scientific communications, it was easy to construct
them from the following very obvious considerations. Like
all physiologists, he naturally began his work under the influence of the schools, and inasmuch as that of Dorpat had,
through its head, Professor Alexander Schmidt, obtained the
ear of all scientific men interested in this subject and dominated
thought on this question, it is plain to see how deeply the views
promulgated from that school occupied his mind at the opening
of his career, gave direction to his researches, and limited the
breadth of his conclusions. On the other hand, it is more
interesting to see how, owing to his remarkable powers of
philosophic inspection, he was enabled to cast off the fetters of
such teaching and to strike out for himself a path of unexampled
originality and truth.

Class I

THE CHEMISTRY OF THE CORPUSCLES OF THE BLOOD AND
THEIR RELATION TO THE PROCESS OF COAGULATION

Taking the first class of his observations, we find that this
includes no less than five papers. In his first paper of all, namely,
' The Structure of the Stroma of Red Blood-corpuscles,' he showed
that, although this latter could be separated into certain com-

ponents—cholesterin, lecithin, paraglobulin, a nuclein-like body, lime salts and iron salts—yet that these various substances were in intimate chemical combination the one with the other, and that the so-called separation of them entailed by its disruption of such combination something more than mere separation of component factors. The experience he gained in dealing with the corpuscular elements of the blood led him to attempt to elaborate the facts upon which the then existing views as to the importance of the leucocyte in the fibrinous coagulation of the blood were based. The first practical assistance to such investigations it seemed to him should consist in more accurate quantitative estimation of the various factors the concurrence of which led, according to the received school view, to the formation of fibrin, and inasmuch as he was then still under the influence of the Dorpat teaching, he set to work to find out how far it was possible in any given specimen to quantitatively estimate what, after the teaching of Schmidt, were considered the active elements in the process of coagulation—namely, the leucocytes. By treating blood first with a 'half-saturated'[1] solution of sulphate of magnesia to preserve its fluidity, then with ether to ensure its being so partly destroyed as to become laky, he finally separated the elements by the centrifuge. Reference must be made to the paper for the discussion of the points raised by him as to how far the results of such observations correctly represent the quantitative estimate of the corpuscles. In the course of this research he paid much attention to the apparent influence of 'leucocytes' when isolated from lymph-glands (therefore more truly lymph-corpuscles) in the production of fibrin, and following up this point with the preconceived ideas upon which he had been trained, he was led to construct the next paper, which is entitled 'The Conversion of Colourless Corpuscles into Fibrin.' The doctrine that is embodied in such a title

[1] *I.e.* A saturated solution to which an equal volume of water is added.

is nothing more than that of Alexander Schmidt, namely, that the colourless blood-corpuscles are active in bringing about the coagulation of the blood, not merely by virtue of their destruction producing ferment, but also by their containing primarily the fibrinogen required, a doctrine which in its most extreme form is embodied in the very latest work of its originator.[1] To determine what actually occurs when leucocytes are brought into contact with plasma, he proceeded necessarily on the two great lines of (*a*) action of leucocytes on plasma separated from the body—*i.e.* shed blood; (*b*) action of leucocytes on living intravascular plasma—in other words, the circulating blood. The interaction which occurs between leucocytes isolated from but recently dead lymph-glands and plasma separated from shed blood may obviously be studied in plasmata obtained in various ways—*e.g.* by cooling and subsidence, by dilution with 5 per cent. salt solution, or by the action of peptone, and in each of the last two cases with the assistance of the centrifuge. Wooldridge found that peptone-plasma previously cooled, although, as (we shall subsequently see why) he discovered, it would not then undergo spontaneous clotting, and would not clot on the passage (to saturation) of carbonic acid, would nevertheless form a dense fibrous coagulum when leucocytes were added to it; and, what is much more important, as throwing light on the interaction of such substances, he found that if this apparently complete clot were removed and that if more leucocytes were added to the remaining serum, clotting would again occur. It is in the light of subsequent events very interesting to remark how he spoke at that time of the action of the blood-plasma upon the leucocytes, as though the fibrin formed was the direct outcome of the breaking up of the leucocytes. Fortunately the paramount

---

[1] ' Ueber den flüssigen Zustand des Blutes im Organismus,' *Centralblatt für Physiologie*, Bd. iv., No. 9, 1890.

desire to keep as far as he possibly could to the conditions of
life led him to follow the research on the second line, namely,
the behaviour of the leucocytes when in contact with living
plasma. By simple and demonstrative experiments of introducing leucocytes into either a localised portion of the living
blood by adopting the Listerian method of suitable isolation of
the jugular vein or directly into the general circulation, he
proved that there was no positive interaction between the
leucocytes and the plasma—in short, no clotting ; and consequently saw that there was something radically erroneous in
the accepted view of the subject, and especially in our mode of
regarding the facts of experiment.

Class II

THE DISCOVERY AND DEMONSTRATION OF THE OCCURRENCE
OF COAGULATION WITHOUT FERMENT ACTION

This becomes still more evident in the next paper, entitled
' Further Observations on the Coagulation of the Blood,' from
which it is obvious that, although there lingered in his mind the
notion that the breaking up of the leucocytes was before all
things important, he looked rather to the changes in the plasma
as the real chemical metamorphosis which ends in the production of fibrin ; for he speaks of his experiments as tending
to show that the blood-plasma liberates from the leucocytes a
certain substance or substances capable of bringing about coagulation of the fibrinogen known to be present in the plasma.
Now the question which naturally arises from such deductions
is,—what substance is conceivably so liberated which has this
active influence on coagulation ? By extracting leucocytes with
alcohol, and by dissolving the alcoholic residue in ether, he
showed that, while the alcohol-ether extract contained no fibrin
ferment (one proof of this position being that heating it to

100° Cent. in water did not in any way destroy its action), the coagulation-inducing substance principally present in the extract was lecithin, and that this, when emulsified with a little sodium carbonate solution, very readily brought about coagulation if added to peptone-plasma slightly acidified with carbonic acid. Finally, his preconceived school notions of the action of the leucocytes received their death-blow when, as detailed in his paper, he discovered that, although leucocytes caused peptone-plasma to coagulate, yet serum which contained large quantities of Schmidt's ferment was inefficient, just in the same way as it is impotent to induce clotting when injected into the living circulating blood. In fact, at the commencement of 1884 he had discovered that, although, doubtless, clotting was frequently brought about by Schmidt's ferment in fluids containing fibrinogenous substances, there were very common—in fact, fundamentally important—conditions of the blood-plasma in which coagulation occurred under circumstances such as precluded the possibility of either the presence or action of a ferment. In short, he recognised and discovered the great principle that there might be *coagulation of the fibrinogen in the plasma of the blood without the intervention of a ferment at all.*

CLASS III

THE TRUE NATURE OF FIBRINOGENS

THE first separate communication to the Royal Society was not made by Wooldridge until this epoch of his career, and we find that the beginning of a new era of extended views of coagulation commences with his brief communication on the ' Origin of the Fibrin Ferment,' and his experiments contained in this communication showed clearly that (1) the fibrin ferment does not exist in normal plasma, that (2) if peptone-plasma be caused to coagulate by treatment with carbonic acid it will then be found to contain ferment; and since peptone-plasma is perfectly free from all corpuscular elements it follows that, contrary to the Dorpat dogma, fibrin ferment is not solely a product of corpuscles, but may arise as a consequence of coagulation changes in the liquid plasma itself. It was this study which led him to the remarkable discovery in peptone-plasma of the fibrinogenous substance which he termed ' A-fibrinogen.' The existence of this body he demonstrated by cooling corpuscle-free peptone-plasma to zero and discovering that this simple physical change produced a flocculent precipitate. If this fibrino-genous substance precipitated by cold be removed, the peptone-plasma will not clot on simple acidification with carbonic acid (*vide supra*). The substance separated by such precipitation (which was termed by him for simplicity 'A-fibrinogen'), on standing, gradually becomes converted into fibrin.

This observation marks a stage in the knowledge of the clotting process for several reasons. Of these the first is the fact that, from the moment when Wooldridge discovered that one of the fibrinogenous substances in the plasma was precipitated by simple cold, it was obvious that the practice introduced by Schmidt of employing in most experiments a plasma obtained by cooling horse's blood at zero, and allowing the corpuscles to subside, was open to the objection that such plasma was in part deprived of a certain proportion of fibrinogen. Indeed, the granular matter described by Schmidt as precipitated between the leucocytes in the middle layer in the tube, Wooldridge pointed out as consisting in all probability of A-fibrinogen. At least it is now quite clear that cooled plasma cannot be regarded as a ' native' plasma if deductions are to be made from experiments in which it is employed.

The second point is the fact observed by Wooldridge, that

the granular precipitate of Λ-fibrinogen, when microscopically examined, was found by him to consist of discoidal elements exactly resembling blood-plates.   The bearing of this on modern morphological research needs no insistence here, and, in fact, is of necessity dealt with in Part II. of this Introduction.   I may therefore leave the matter, only remarking that Wooldridge, in drawing attention to this fact, always laid stress on the discoidal and occasionally globular character of the particles of this precipitate as evincing a kind of pseudo-crystallisation of proteid from the plasma.   (See also Part II.)

A third and very important reason for assigning special notice to this observation of the effect of cooling is the fact that it is the first pure case of a precipitate being obtained from corpuscle-free blood-plasma in which the precipitated substance without further treatment yielded amorphous fibrin.

At the conclusion of this note he put forward in so many words the important issue ' that fibrin may be deposited from blood by simple physical means without any ferment process.' To pursue the further analysis of plasma he next examined the condition of peptone-plasma after the removal of Λ-fibrinogen, and he discovered that such plasma still contained coagulable matter—in other words, fibrinogenous substance—which would not form fibrin upon the simple addition of fibrin ferment derived from serum, but would give a clot if treatment with carbonic acid was also resorted to.   His observations also went to show that the coagulable body existed in the form of a precursor of fibrinogen, and not as a true fibrinogen, in a majority of cases ; but that, in addition, fibrinogen, as it is ordinarily understood (i.e. Hammarsten's), was frequently present in the peptone-plasma, and was demonstrable as such by the formation of an abundant clot on the addition of ferment.

The last paper of this, the third division of his work, namely, ' The True Nature of Fibrinogens,' is the one which embodies his most striking and far-reaching discovery—namely, the production of extensive intravascular coagulation in the living organism. Always striving to arrive at the chemical conditions of living plasma, he was led, by virtue of his observations on the relation between lecithin and proteid, to see what could be obtained in the shape of a lecithin-proteid compound from such organs as were known already to contain a notable proportion of the former constituent. Consistent to his view that the soundest work would be obtained by employing as simple as possible an extract of the tissue in question, he prepared a solution by extracting the finely divided organ with water, and precipitated from this watery solution a lecithin-holding proteid by means of acetic acid ; and having obtained in this manner, as he himself recognised, a complex mass, he redissolved it in dilute alkaline solution, and employed that solution in his experiments. The injection into a vein of this solution, which has no mechanical properties that could induce coagulation, as was once absurdly suggested (see Appendix, p. 305), produced instantly clotting of the living blood ; the phenomenon which so many workers had attempted to obtain by the employment of defibrinated blood, solution of ferment, hæmoglobin, leucocytes, extractives, &c., but which no one had succeeded in attaining except in a most imperfect and inconstant manner. This discovery is of far wider importance even than the light which it throws on the obscure subject of Thrombosis intra Vitam. It is of fundamental value in the great question of the chemistry of clotting, because in this discovery we have for the first time the solution of a proteid which is capable of so vastly altering the substances precursory of the fibrinogens, as well as the fibrinogens themselves, in the living blood that it can cause firm fibrinous clot to result from the interaction caused by its presence. Further, the solution not only contains no fibrin ferment, but also, as we have seen before, injection of large quantities of fibrin ferment have

practically no such action ; and finally, when the proteid solution
obtained from the thymus has produced its remarkable clotting
influence, only a trace of ferment can be detected in the altered
blood.   In accordance with his previous methods of research, he
turned to the next point—namely, the concurrent existence and
action of lecithin—and he then discovered that if, from the complex solution, the lecithin present were extracted by alcohol and
ether, the proteid thus left, while as before still soluble in dilute
alkalies, no longer possessed the power of causing intravascular
clotting.   Of course, as we shall see, Wooldridge was keenly
alive to the probable metamorphosis of the proteid by the treatment of the substance with alcohol, &c., for the purpose of the
extraction.   As he had previously found that lecithin alone, if
injected into vessels, had no influence in producing intravascular
clotting, he believed that the fibrin-forming influence of the
solution of 'tissue-fibrinogen' was due to some fundamental
chemical process of interaction between proteids, and that that
interaction was greatly favoured by the concomitant presence
of lecithin, and he never thought, as has been supposed, that
lecithin *per se* was capable of converting fibrinogen into fibrin.

This remarkable paper,[1] therefore, as a matter of fact,
embodies a principal demonstration of his discovery that fibrin
formation depends, at least in its most striking form, on interaction between proteids, and not upon fibrin ferment.   It is
curious that such a discovery as this should not have secured for
his work in his own country the reputation and sympathy which
it gained him abroad, and I can only suggest that a curious
intellectual blindness must have prevented some from seeing
that, since at least we could not be supposed to know as yet
more than the mere surface of such an admittedly complex and
difficult subject as that of the Coagulation of the Blood, it was

[1] For the bearing of this discovery on the pathological process of thrombosis, see the second part of this Introduction.

most illogical and unscientific to assume that there was only one mode of fibrin formation, and that this was by the action of Schmidt's ferment. Yet this, nevertheless, was done by his critics; and now, when we come to consider the summary of his work, given to the Royal Society as the Croonian Lecture, it is certainly a matter of no honour to those in whose hands the determination of such matters is put that this posthumous collection of his works should be the first opportunity which physiologists and pathologists can have of hearing his side of the question, as stated by himself in this important communication in the year 1886.

CLASS IV

THE CROONIAN LECTURE

I HAVE placed the Croonian Lecture alone in a separate class of his papers, inasmuch as it speaks for itself in a singularly clear way of the change which had come over his scientific trains of thought, and which had consequently led him by logical steps from one discovery to another. I shall not intervene between it and the reader more than to point out what is now perhaps rendered clearer by the process of time, that in this work he describes for the first time in detail the truly scientific notion which we have seen from his previous papers was in process of intellectual development, viz. that the formation of fibrin is a chemical process due to the interaction of one proteid, so-called fibrinogenous substance, upon another, that this process is analogous to the coagulation changes which are commonly known and familiarly spoken of as 'rigor mortis,' and that the conversion of the liquid fibrinogens (present in solution in blood-plasma) into the solid amorphous substance fibrin is not of necessity due to the presence of a ferment, but may, and probably does, occur as a simple death phenomenon,

C

just as in other living tissues. In developing this idea he sets forth in the Croonian Lecture a division or classification of the fibrinogenous substances in the plasma of the blood which, however, he did not seek to employ more than as a working hypothesis, although, as a matter of fact, the individual items of it were all grounded upon and established by the experiments embodied in his researches of the previous five years. In this classification or separation he denoted, for simplicity of reference in discussion and to avoid dogma, the various fibrinogenous substances which could be distinguished by rough tests from each other by terming them respectively A, B, C, D Fibrinogen. As a matter of fact he only separated three before his death, and even then always insisted that such separation was of necessity artificial, and that any of the further members of such a series might well be nothing but a metamorphosis of an earlier member (in fact, its immediate predecessor), or the resultant of an interaction between any two other members of the series.

CLASS V

FURTHER ELABORATION OF HIS OBSERVATIONS ON COAGULATION

BETWEEN the reading of the Croonian Lecture, on April 8, 1886, and the furnishing, on November 4, 1888, of his first report to the Scientific Committee of the Grocers' Company, Wooldridge produced several papers and lectures, in which his views were elaborated and supported by numerous fresh experiments, which amplify and thoroughly confirm in every particular the conclusions which he came to in the Croonian Lecture. The two most important of these papers are, first, that which he contributed to the Festschrift of Professor Ludwig in 1887; and, second, the contribution to the coagulation question which was published in the first half of 1888 in the ' Archiv für Anatomie und Physiologie.' Full of the view that the substance commonly spoken of as fibrinogen, whether isolated by the method of Hammarsten or that of Schmidt, is, by virtue of its mode of preparation, not existent in its isolated form in normal—*i.e.* living—plasma, he examined the relations, chemical and physical, which surround fibrinogen as thus known, and he then demonstrated that it could not pre-exist in the plasma, since solutions of it are coagulated by fibrin ferment, while, as before repeatedly stated, fibrin ferment does not cause any coagulation in normal living plasma.

CLASS VI

FIRST REPORT TO THE SCIENTIFIC COMMITTEE OF THE GROCERS' COMPANY

As in the case of the Croonian Lecture, I have thought it right to place the First Report to the Scientific Committee of the Grocers' Company in a position by itself [1] among his other works; and I am inclined to do so for several reasons, especially since it is probably not generally known that the same unjust hostility which suppressed the publication of the Croonian Lecture prevented this Report from being widely circulated and known, although it does not contain a single fact but what is based upon experiments which had extended over seven years, and had been repeatedly demonstrated by him, and witnessed by very many of his fellow-workers and teachers.

In introducing *seriatim* his previous communications, I have abstained from referring in much detail to the most important observation which he made in 1884 in the paper which he communicated to the Royal Society, and which contained the discovery of A-fibrinogen. In that paper he mentions that the

[1] The translation into German by Prof. Max von Frey of this Report is a very suitable monument of his work in the country which showed him the most genuine scientific appreciation.

precipitate produced by cooling, when examined microscopically, consisted of corpusculate particles precisely resembling in every particular the blood-plates described by Bizzozero, Löwit and others. Like Löwit, he regarded these bodies purely in the nature of a precipitate, but pointed out that their morphology had a certain significance, inasmuch as their circular or discoidal form was strongly suggestive of the shapes adopted by various albuminous compounds when in a crystalline or pseudo-crystalline state, and he looked upon the morphology of this precipitate as suggestive of the changes which were intermediate between the physical condition of the liquid fibrinogen and the solid fibrin. In the Report to the Grocers' Company he not only states most clearly his position with regard to this question, but he again examines the effect of leucocytes upon the normal living plasma, and he dwells at some length on the experiments which were brought forward by the Dorpat School with the object of maintaining the old and stereotyped view taken by Schmidt, that the fundamental process in coagulation is the active participation of the white corpuscles. Examination of this, the first part of the Report, exhibits most clearly the scientific spirit in which he always treated critically a debateable point of so great importance. In the second part of the Report he proceeds to elaborate the position that I have already shown he was the first to discover and to advance as a new experimentally determined explanation of fibrin formation in the coagulation of the blood, namely, the Interaction of Fibrinogens.[1] The whole of his work which bears

---

[1] This is the fitting place to refer again to the use of the term ' fibrinogen,' and especially in connection with recent researches on the concurrence of certain salts in the process of coagulation. Wooldridge always contended that every fibrinogen separated as stated in physiological text-books was of necessity a highly complex body, and that a fibrinogen in the living plasma, and as obtainable from such a plasma as peptone-plasma, might conceivably have a percentage composition as follows : 94 per cent. albumen, 4 or 5 per cent. lecithin, 2 or 1 per cent. salts, the latter containing calcium. The

upon this point is alluded to in this part of his Report, and is dealt with in a brief manner, because he always determined to make the question of the interaction of the fibrinogens the feature of a second Report, which was due to the same scientifically enterprising Company. Unfortunately for science, he had only placed together in a fragmentary fashion the groundwork of this second Report when his lamentably premature death occurred. It was in his mind to represent sufficient of the history of the subject to show what he often suggested in conversation, namely, that Schmidt unconsciously admitted the necessity of viewing the phenomenon of coagulation as a chemical interaction between proteid fibrinogens when he introduced paraglobulin or fibrinoplastin as the third necessary factor in the process of clotting. On this point one experimental fact, always so striking to the unbiassed critic, is alluded to in this fragment as having been demonstrated by Hammarsten, and which is unquestionably absolutely fatal to Schmidt's view of the activity and importance of the various substances taking part in this great chemical change. I refer to the fact, pointed out by Hammarsten, that since the fibrin formed was quantitatively less than even the fibrinogen present,

intimate correlation between certain salts and various albumen constituents of the tissues has often been discussed. The researches of Ringer for many years showed the extreme importance of calcium salts to the functional activity of living protoplasm, but the influence of calcium in the phenomenon of conversion of living plasma into clot was not investigated until after a suggestive observation by Hammarsten; the subject was then taken up by Green, later by Ringer and Sainsbury, and finally by Arthus and Pagés. These latter authors have found that if all the calcium compounds present in plasma be rendered insoluble in the form of oxalates by receiving blood into a solution of oxalate of potassium, such decalcified blood remains fluid. Restoration of a soluble calcium salt restores the power of clotting. It was shown by Green that there was no combination between a 'zymogen' and calcium sulphate, with the result of production of fibrin ferment, but Pekelharing suggests that the action of fibrin ferment may be such as would bring about the interaction of a globulin and lime salts to the production of fibrin. The efficacy of calcium being admitted, the point cannot further be discussed here, as being as yet in such an indeterminate stage.

it is clear that fibrinoplastin could add nothing in the sense in which it was supposed to by Schmidt. Wooldridge goes further in the second fragment, and suggests that Schmidt, from *a priori* grounds, could not bring himself to believe that a proteid interaction could occur, and therefore fell back entirely on the presence of the ferment as a necessary component.

Quite recently[1] Professor Schmidt has put forward his most recent views in a preliminary communication on the fluid condition of the blood in the (living) organism. In this he states that fibrin ferment arises out of all forms of protoplasm as a result of the catalytic action of plasma. Ignoring the fact that Wooldridge had demonstrated that the formation of fibrin may occur in the total absence of fibrin ferment, and, further, that the work of his (Schmidt's) pupils, Rauschenbach, Grohmann, v. Samson and Nauck, is not conclusive, he lays down as a law, but quite without foundation, that fibrinous coagulation is a function of the cell. He also attributes to von Samson and Nauck the discovery that various extractives, including lecithin, act upon filtered horse-plasma, leading to the production of fibrinous coagulation, quite regardless of the fact that, years before, Wooldridge had not only demonstrated the action of lecithin on corpuscle-free plasma, but written a great deal on the bearing of its influence. After having laid great stress on his now exclusive view that clotting is entirely dependent on the breaking up of cell protoplasm, Schmidt contributes to the subject the statement of some experiments that he has been carrying out on masses of various forms of protoplasm— *e.g.* blood-corpuscles, lymph-cells, spleen-cells, liver-cells, &c. These he has extracted with alcohol, and then treated the remaining mass with water, and both the mass and the watery solution of the same he found, he states, to be productive of an

---

[1] *Centralblatt für Physiologic. Verhandlungen des Internation. Congresses zu Berlin*, 1890.

inhibitory effect on the power of filtered horse-plasma to clot. He comes to the remarkable conclusion that the fluid condition of the blood in the organism is, therefore, a function of the cells, that inside the living body cells prevent clotting, but that outside they determine it. Of course, Wooldridge showed, as long ago as 1882, that leucocytes, when introduced into the circulating blood, not only produced no clotting, but that the blood drawn immediately showed little tendency to clot; but he did not, like Schmidt, immediately conclude that the leucocytes were exerting an inhibitory influence on the development of ferment; and he similarly showed that when leucocytes were added to shed blood they hastened its clotting in a very marked way. It is very difficult to follow the ideas which can have induced Schmidt to formulate these very absolute statements on what is transparently insufficient ground, for they amount to nothing more than this : that the leucocytes, which, according to him, are the sources of ferment, and which also, according to him, are broken up by plasma, nevertheless, when introduced into that fluid while in circulation, equally possess, according to him, the remarkable property of preventing themselves from being broken up to the production of ferment. Not only has he been thus brought to an almost inconceivable position, but his statement is obviously incompatible with the facts of intravital and intravascular clotting—e.g. thrombosis—since that, according to the view of Schmidt, can only be due to the breaking up of leucocytes, such breaking up being the result of the action of the plasma ; and yet, when leucocytes are introduced for this purpose into the circulation, they do not produce any thrombosis. The other conclusions to which Schmidt is led in this communication are equally remarkable and equally independent of what has been done in this field by other workers. I need not mention, therefore, more than the last, in which he makes the assumption that fibrin is solely a derivative of the cells of the body, that it is in all cases an amorphous separated-out cell substance, regardless of the fact of its having been obtained from corpuscle-free plasma.

It is a great pity that such an interesting and important subject as this should certainly be made no clearer, but only clouded, by observations and conclusions in which the work of other physiologists is disregarded, and the conclusions drawn are based upon preconceived notions rather than a critical survey of actual facts.

Few points exhibit this more clearly than a comparison of Schmidt's recent *dicta* on the inhibition of coagulation and the previous observations on the delayed clotting made by Wooldridge, prominently formulated in this First Report of 1888, and fully described by him in 1886.

Wooldridge showed, by the intravascular injection of tissue-fibrinogen, that two opposite phases of clotting of the intra-vascular blood occurred simultaneously:

(*a*) The blood in part clotted immediately;

(*b*) The blood in part remained fluid and non-coagulable (*i.e.* by ordinary means);

therefore he called (*a*) the *positive* phase and (*b*) the *negative* phase of coagulation.

What is of more fundamental import is his discovery that the development of these two phases proceeds in quantitative parallelism with the amount of 'tissue-fibrinogen' injected. The fact of the remainder of the blood being, as a result of the thrombosis, non-coagulable is, of course, of far-reaching significance, and as such was fully realised by Wooldridge. Looking at the phenomenon from his point of view, it revealed a very profound alteration in the chemical constitution of the blood, and suggested to him a crowd of possibilities, the most valuable of which was unquestionably his idea of altering by this means the constitution of the blood, so that it should no longer, to use his words, be a suitable (sensitive) pabulum or field for microbic infection (*see* Pathology, p. 28).

To such generalisations of breadth and worth was he led by his examination and discoveries of the negative phase of clotting as evoked by intravascular injection of fibrinogens.

Although it may be disadvantageous to institute comparisons, it cannot but be admitted that the recent complicated hypotheses of Schmidt have borne no such fruit.

The rest of this valuable Report deals with the nature of fibrinogen, the change which not only A-fibrinogen, but the correlative compounds present undergo when treated with strong reagents and processes, which cannot but greatly alter them; and from that position he proceeds to point out what the inter-action of proteid fibrinogens means scientifically speaking, and how, like other chemical reactions, the phenomenon is dependent on the influence of the mass of any one of the component sub-stances.

Class VII

FURTHER PAPERS, INCLUDING THE UNFINISHED SECOND REPORT TO THE GROCERS' COMPANY, DEALING WITH THE QUESTION OF THE INTERACTION OF FIBRINOGENS AS THE CAUSATIVE FACTOR IN FIBRIN FORMATION.

SUFFICIENT has been said to explain the groundwork of Wool-dridge's conception of the process of coagulation, and his view of it as the outcome of two chemical bodies of the same class, but different compositions, interacting upon each other. Study of all his experimental work shows how this view is not only in harmony with the theories of modern chemistry, but is remark-ably explanatory of all experimental researches on the subject, whereas the ferment theory of Schmidt is inapplicable to a large number, especially those of intravascular coagulation.

The papers in the present class are extremely interesting to study from this point of view, and especially the unfinished second Report to the Committee of the Grocers' Company. Although fragmentary, and in part disconnected, it is highly suggestive, and cannot fail to strike all interested in helping forward this incomparably important question of physiology as presenting an original solution of the problem in a clear and convincing form.

Further description of these, the last expressions of his intellectual energy, is unnecessary.

Summary of Wooldridge's Views of the Causation of Clotting

To summarise Wooldridge's views of coagulation nothing is easier, inasmuch as he always held that clotting could conceivably occur in one of two different ways:

(*a*) By the action of fibrin ferment upon a fibrinogen, *i.e.* an albuminous substance which could be converted from a fluid state into that of the amorphous solid which is commonly known as ' fibrin.'

(*b*) That fibrin is the result of an interaction between two or more fluid fibrinogens, the chemical composition of which should differ sufficiently to permit of interchange occurring between them.

The former of these two views was formulated by Schmidt; it was Wooldridge's merit that he discovered that fibrin could be produced in the total absence of fibrin ferment; in short, he discovered the second manner in which fibrin can conceivably be formed.

PART II

I HAVE next to deal with those of his papers which treat essentially of what are commonly called pathological processes. Of these, unhappily, we have but very few. Fortunately, however, they embody the great discovery that it is possible to arrest the pathogenic action of microbes by injecting into the system certain albuminoid solutions and so altering the constitution of the blood, &c.—the first discovery of a genuine process of immunity evoked by method other than that introduced by Pasteur, namely, the attenuation of a virulent organism. I will, therefore, place the communications dealing with protective inoculation first in a class by themselves.

The second paper is one upon the results obtained by the method he discovered of producing intravascular clotting with absolute certainty and to a known degree. While dealing with this particular paper I shall discuss the bearing of his discovery on the great pathological subject of Thrombosis, and therefore place this paper in a second class. Lastly must be placed in a third class by itself the highly suggestive paper he wrote upon Auto-Infection in Cardiac Disease.

CLASS I

PROTECTION AGAINST THE ACTION OF PATHOGENIC ORGANISMS

ALWAYS consistent to his view that the first consideration in planning any investigation of a chemical nature must be the attempt, so far as is possible, to deal with substances which most closely resemble the living tissue, Wooldridge approached what he often spoke of as the culminating deduction to be made from his work—namely, the direct application of the general physiological principles discovered to the elucidation of disease processes—with the determination to discover what would be the relation of the growth of a well-known organism like that of anthrax to the chemical substances which more nearly resembled living tissue than any other, namely, this class of tissue-fibrinogens. It not being possible, as he pointed out, to obtain an artificial solution of a tissue, the nearest thing that can be done is to make an extract of a tissue, and this he proceeded to do as described before in the preparation of tissue-fibrinogens. His earlier ideas were clearly that as the modified anthrax virus, forming the so-called vaccine of Pasteur, must be supposed to effect a sudden change of a chemical nature in the tissues, so Wooldridge thought that, by growing the anthrax bacillus in a solution of tissue-fibrinogen, he might so alter that substance or substances and produce therefrom, as the result of metabolism, products which might have the power of conferring immunity. After many attempts and elaborate variation of the conditions of his investigation, he discovered the great principle that the immunity obtained depended far more upon chemical alterations of a tissue-fibrinogen during its preparation, than upon the influence exerted by the growing bacillus. From this position it was an easy step for him to attempt to obtain the tissue-fibrinogen in that particular condition which would bring about the desired immune state, and he then discovered that, if a solution of a tissue-fibrinogen was partly (principally) coagulated by boiling, and then coarsely filtered, so that some of the finely suspended coagulum formed part of a protective material, this was the potent factor which produced the protected state when it was injected into an animal. This observation records the discovery that immunity can be obtained, not, as had previously been supposed, only by the inoculation of attenuated micro-organisms or their products, but by the administration or introduction into the system of a chemical substance which had never come into relation with or was in no sense a product of the life of a micro-organism. Wooldridge was thus the

discoverer of ' defensive proteids.' Such an advance in our knowledge of the pathology of infective diseases and the means of
counteracting them cannot be placed other than among the most
· remarkable in the range of pathology, and it shows what
an extraordinarily prolific subject pathology is when treated
from the point of view taken by a genius of the first rank,
for the whole substratum of this discovery was the method
which ran through all Wooldridge's work—namely, his great
principle of so arranging research chemically in physiology
as to produce as little artificial change in the tissues as possible.
Since he published this discovery the subject of immunity has
been handled in various ways, both morphologically and chemically. A few words, therefore, will not be out of place to point
out how these investigations have confirmed or antagonised
the principle embodied in Wooldridge's discovery. A brief
sketch is rendered the more easy in this matter, inasmuch as
the different observers whose work is devoted to this investigation have now, after much discussion, stated their views cate-
gorically. Without entering into unnecessary detail, it is clear
that in a certain number (the large majority) of infective diseases
phagocytosis occurs, whereas in a few—e.g. diphtheria, tetanus—
it is scarcely present. Further, that the blood of animals, which
have been rendered immune against infection by culture of a
virulent organism in consequence of a previous injection of the
same in an attenuated state, contains some chemical substance
the nature of which is as yet obscure, but which, if injected into
the blood of another animal, will confer the property of immunity
on the subject of such treatment. In this connection it has to
be mentioned that the intervention of an inorganic chemical
substance—e.g. the trichloride of iodine—will so alter the blood,
and probably the tissue fluids, that the same when passed into
another animal will similarly confer immunity on the second
individual. Still more extraordinary are the experiments of

Ehrlich, who, by his observation on the toxic albuminoids of the castor-oil plant, has found that not only is an immunity against the chemical action of such substances readily obtained, but that such immunity must mean a chemical metamorphosis of the tissue juices, since the descendants through the female subjects rendered immune are also immune; whereas the descendants through the male immune subjects possess no such favourable characteristic. An exact parallel, of course, to this transmission by heredity on such lines is found to be present in various disease conditions, but the deduction to be made from such facts is that the question, in a certain number of cases at any rate, is a matter affecting primarily the chemical condition of the tissue juices. Whether this primary factor is followed by, and finds scope for its agency in, an influence which it brings to bear on phagocytes, as has been suggested, is a point which has no direct bearing upon the present matter. The general deduction which I will venture to draw from a survey of these more recent researches is that they unconsciously reflect the idea conveyed in Wooldridge's great paper on ' Protection,' since, as has already been stated, in that paper, the discovery that it was possible to obtain immunity by injecting into the circulation an albuminous fluid of a special nature, but quite independent of the action of micro-organisms, was in effect a presage of all subsequent researches in that it obviously afforded the first example of alteration of tissue juices. One cannot help regretting that the width of Wooldridge's views was not recognised, that the great principles which underlay his observations were apparently not grasped by those who followed him, so that we have not got, since his untimely death, a continuation of the work exactly on the lines on which he commenced it. As far, however, as his scientific memory is concerned, nothing can be more gratifying than to see the way in which the truth of the great principles which he enunciated has been established by subsequent research.

The only work which has been directly carried on on the line indicated by Wooldridge is that by Wright (*Brit. Med. Journ.*, September 19, 1891), who found, upon repeating Wooldridge's experiments exactly, he obtained complete confirmation of Wooldridge's statements. As regards the question as to what classes of bodies are particularly contained in the substances termed ' tissue-fibrinogen,' Wright came to the conclusion that they belonged to the class of proteids which Hammarsten termed ' nucleo-albumens.' In his investigations Wright took special precautions to prevent the possibilities of the tissue fibrinogens employed containing bacterial products, or possibly other organisms than anthrax, when that was grown in the fibrinogen solutions.

CLASS II

I HAVE already in Part I. referred in the proper place to his discovery that if certain tissues—*e.g.* testis, thymus, &c.—were (under circumstances which precluded the intervention of decomposition) extracted with water, and if the watery solution thus obtained of proteids with correlative salts and other substances—*e.g.* lecithin—was then precipitated with acetic acid, and the precipitate separated and dissolved in dilute saline solution, that injection of that solution produced widespread intravascular clotting of the living blood. To gain an insight into the mechanism of a process whereby the living blood clots within the blood-vessels as a consequence of disease, numerous valuable experiments have been made, especially by the pupils of Schmidt. Taking the view of the Dorpat School first, that clotting is simply and solely due to the liberation of ferment, and fibrinogen as well, from the broken-down protoplasm of the corpuscles, it should follow as a logical necessity that the introduction of ferment into the blood ought to produce wide-

spread and instant clotting; but, as a matter of fact, it does
nothing whatever of the kind: fibrin ferment injected into
the blood-stream does not produce intravascular clotting or
thrombosis.   Neither, as Wooldridge first showed, and as
Rauschenbach afterwards found, does the introduction of leucocytes to a large amount evoke the result, although, of course,
according to the Dorpat hypothesis, they ought inevitably
to produce this effect.   Reference may here be made to page
22, whereon is stated the last attempt of Schmidt to endeavour
to explain away a fact so damaging to his position.   I need
hardly remind those interested in the question that the oft-
quoted experiments of Kohler on the one hand and Krüger on
the other do not in the least invalidate these statements as
regarding the inefficacy of both ferments and leucocytes to
produce clotting in the living blood, since in all the experiments
with defibrinated blood, and with what Kohler called 'ferment-
rich' blood, the fluids injected do not merely of course contain
ferment, but also enormous quantities of the stromata of red
blood-corpuscles, which, as Wooldridge showed, are perfectly
capable of inducing a certain degree of intravascular clotting, a
point which I shall refer to again directly.   While, finally,
as regards Krüger's experiments with leucocytes, the inaccuracy
of his method, by which he injected quantities of albuminous
substances in solution besides leucocytes, is dealt with in
Wooldridge's first Report to the Grocers' Company.

To return now to Wooldridge's discovery: it must first be
observed that he did not merely find that a certain albuminous
solution would cause the intravascular—i.e. living—blood to clot,
but he discovered

(a) That certain predisposing conditions of the blood-plasma
favoured this clotting.

(b) That it began in a particular part of the circulatory
system.

(c) That, other things being equal, the extent of the clotting was proportional to the amount of the solid albuminous substance which was in the solution injected, and which he termed 'tissue-fibrinogen.'

(d) That if from this albuminous solution the lecithin was all carefully extracted, the residue, although still soluble in dilute alkaline solutions, was incapable of inducing this phenomenon.

I will briefly consider the above-mentioned points, and show to what important general conclusions he was led by these discoveries. In the first place, it will be found that, both in his paper and in his first Report to the Grocers' Company, he showed that to successfully induce intravascular clotting the most important predisposing factor was the condition of the blood, that is to say, of the plasma. He very soon discovered that the diet which should result in a condition of blood most favourable to intravital clotting was one that must contain fat. This point alone suggests that the fundamental change in fibrin formation must be one of purely chemical nature, and that it is probably not one directly connected necessarily with the destruction of protoplasm. At this point there arises a question which I have also already partly alluded to in the first part of this Introduction. I refer to the observations of all who have directly investigated the subject of intravascular coagulation, namely, that within the blood-vessels, when these latter have been exposed to injurious conditions, blood-plates make their appearance in the plasma at the seat of injury, and the first clot that is formed is truly a plasma-clot, the only apparently corpuscular elements in it being these blood-plates and a few entangled leucocytes. Exactly the same observation is made when the structure of the blood is mechanically interfered with—e.g. by a foreign body, such as a glass fibre, &c. Or again the same change, namely, the formation (collection, according to

Bizzozero) of a mass of blood-plates, is seen to occur in the
plasma when the vessel wall is punctured and the blood oozes
forth and is exposed to the air. At the present time judgment
is withheld by practically all authors as to whether blood-plates
are preformed structures existing in the living circulating blood
or whether they are artifacts, the result of the changes in the
physical condition of the blood. Wooldridge always regarded
them as identical to all appearances with the precipitate which
he obtained from peptone-plasma by cooling, and which he
termed A-fibrinogen; consequently he was led to believe
(although he was naturally too scientific to express a dogmatic
opinion) that possibly the two things were identical, and that
possibly the appearances observed by inspection of the circulating blood at the commencement of the thrombotic process were
truly of the nature of a precipitation of a fibrinogen. It will be
remembered, and reference should here be made to page 271,
that in his Arris and Gale Lectures to the Royal College of
Surgeons he developed the view that the corpuscle-like blood-
plates on the one hand and his precipitate of A-fibrinogen
ought both to be regarded as pseudo-crystalline formations of
complex proteids. The discoidal and spherical shape in which
proteid substances are usually crystallised when crystallisable in
conjunction with metallic salts lends considerable colour to this
view regarding the nature and formation of these bodies.
The question has a great importance, because there is no doubt
whatever as to the facts observed from the time of Zahn to that
of Bizzozero, in which the origin and structure of the colourless
blood-plate clot have been thoroughly investigated. The only,
but most important, point at issue is the question of the pre-
existence of blood-plates as a distinctly separate corpuscular
constituent of the blood circulating during the healthy living
condition of the blood. As this is scarcely the place to
elaborate the *pros* and *cons* of such a recondite matter, I will

dismiss it by suggesting that, when Wooldridge's observations are taken in conjunction with the later researches of Löwit, the conclusion appears to me fair that the weight of evidence is in favour of their being artifacts, and consequently that, since they unquestionably are the leading feature in the early stages of the formation of a clot, we are, at the very commencement of this · most important subject of thrombosis, shown that the development of that condition is connected rather with profound alterations in the fluid albuminous constituents of the plasma than to primary changes in the leucocytes. Indeed, a conclusive observation on this matter is that, if any one will repeat the classical experiment of Zahn or of Eberth and Schimmelbusch, they will find that, although there is a well-marked thrombus, nevertheless the leucocytes entangled in it show, even in the well-formed clot, not the least indication of breaking down; in other words, morphologically, the doctrine of Schmidt has in this matter not a single fact to rest upon. We may now profitably turn to the next great point among Wooldridge's discoveries on intravascular clotting. I refer to the topographical distribution in the body of the thrombotic process. He very early discovered that, if but small quantities of tissue-fibrinogen were injected into the venous system in dogs, the albuminous fluid passed through the right side of the heart, through the capillaries of the lungs, through the left side of the heart, and through the arterial system and the mesenteric capillaries into the portal vein before [1] evoking intravascular coagulation ; in short, he

---

[1] Of the various adverse criticisms that were advanced against Wooldridge's researches, perhaps the most absurd and unwarrantable was the assertion that the solution of tissue-fibrinogen was so viscid as to cause clotting by mechanical obstruction of the vessels. Unreasoning hostility could scarcely go further, since, as Wooldridge had shown in his first paper, and as is just stated, the solution passes through *two* sets of the finest *capillaries* before producing its effect. It is as equally regrettable as the criticism was foolish, that whereas Wooldridge's Croonian Lecture was refused publication, criticism of this nature was printed and published *in extenso* by the Royal Society.

discovered that the portal blood was the first to react with the solution to form a clot. Of course the very important bearing of this observation is at once obvious when we consider how very different the chemical constitution of the portal blood is as compared to that in other parts of the body. He next found that if the dose injected were very slight the thrombosis was very partial, and that the clot might soon be broken up again' and redissolved, only slight traces being found of it at the end of some days, but that naturally secondary changes in the liver resulted in proportionate degree to the amount of clotting which had gone on in the vein. To these changes in the liver I will subsequently refer. Supposing now the dosage had been more considerable, he next observed that the clotting process extended right through the liver and invaded the right side of the heart, usually stopping there in the dog, but in the rabbit proceeding to the arterial system, so that in a very marked case the whole of the vessels of the body might be plugged with coagulum. If, as these experiments from the outset tended to show, we are in thrombosis dealing with a simple chemical interaction, we might naturally *a priori* expect that the time duration for the full development of the phenomenon would be extremely short. And, as a matter of fact, it is practically instantaneous. No lecture experiment that I am aware of affords a more remarkable or striking demonstration than this, or is more certainly and easily performed. This instantaneous character of the reaction is a further support to the view that thrombosis is much rather the result of a chemical reaction between albuminous substances in solution than a process dependent upon first a stage of destruction of leucocytes; then a second stage, the evolution of ferment and fibrinogen therefrom; next, a third stage of reaction of ferment upon the fibrinogen in the plasma; and lastly, a fourth stage, the production of amorphous fibrin.

A conclusive proof of the error of imagining that the ferment hypothesis is adequate to explain such a remarkable and extensive pathological change in the blood is shown by the fact that from the clot and blood thus formed practically no ferment can be obtained; hence it follows that from this point of view also the Dorpat teaching fails.

Of all the other facts connected with this process, as first discovered by Wooldridge, nothing is more interesting, chemically speaking, than that the extent of the clotting produced is proportional to the amount of tissue-fibrinogen introduced. A small quantity produces only slight thrombosis in the portal vein and only a temporary disarrangement in the general health, a larger dose evokes febrile symptoms and extensive changes in the liver, while a still greater dose causes immediate death, and a post-mortem immediately performed reveals clotting throughout the whole body. From these facts it is abundantly evident that the phenoménon is dependent upon what in modern chemistry has received the appellation of 'mass-action.' Of course, although very suggestive, this fact does not adduce absolute proof as to whether we are dealing with the destruction of leucocytes as a direct consequence of the injection of this liquid, or, as seems most probable, with direct chemical reaction of the latter upon the fluid fibrinogens in solution in the plasma. The last point to be mentioned in describing the process which results in what he not improperly called the *positive* phase of clotting, is the part played apparently by the correlative substances present—*e.g.* lecithin. The observation by which he found that extraction of the lecithin from the tissue-fibrinogen takes from it the power of inducing intravascular clotting I see no reason for doubting, although other changes and alterations in the fibrinogens (as Wooldridge himself pointed out) may follow as a secondary occurrence due to the direct effect of the extraction process. In any case, however, the acceleratory influence of lecithin, first discovered and demonstrated by Wooldridge, has

subsequently been fully confirmed by the pupils of Schmidt, so that
one need not further discuss this accessory point or doubt the
accuracy of his interpretation of the facts.

So far I have spoken of the successful production of clot
during life within the vessels—*i.e.* the positive phase of intravascular coagulation, using Wooldridge's terminology. There
remains a most important fact for consideration, namely, his
discovery of the *negative* phase of clotting under these circumstances; in fact, the undue preservation of the fluid state of
the blood even when shed. This is noticed when the amount
of tissue-fibrinogen injected is insufficient to cause very widespread clotting. In such an experiment, when after the injection has been made and a well-marked but limited clot formed
in the portal system, if blood be immediately drawn, it will be
found that the normal degree of coagulation does not occur, or
rather is postponed often for as much as twenty-four hours, and
yet if to such still fluid but shed blood leucocytes or fibrinogen be added, clotting occurs. In his Report to the Grocers'
Company he ventured on the suggestion that this change in
the blood is analogous to what occurs in zymotic disease.
Having thus discovered a means whereby the portal vein can
be thrombosed to a degree that is quite at the will of the
experimenter, Wooldridge had as a fact found a method of the
utmost value to pathological science, a fact which he was, of
course, the first to recognise. His plan of tissue-fibrinogen
injection furnished, in fact, the first opportunity of investigating
the condition of hæmorrhagic infarction of the liver, and had
it not been that he was interrupted from following up this
work shortly before he died it would doubtless in his hands have
very speedily furnished pathology with some most valuable conclusions. Two conditions which his procedure evokes immediately attract attention, and were already pointed out by his
keen observation. The first is that, long after the clot has

disappeared from the portal vein, the remaining effects on the liver show themselves as (*a*) necrotic areas, (*b*) islands of cirrhosis. The second great feature is the change effected in the blood which gives a specific character to the condition, and which is evidenced by the occurrence of remarkable transudation of blood into the surrounding tissues; in short, extensive and widespread hæmorrhages, not merely in the liver but elsewhere. In fact, to use his own words, there is ' set up a temporary and mild sort of hæmorrhagic diathesis.'

CLASS III

ON THE PATHOLOGY OF CARDIAC DISEASE

ONE of the most remarkable examples of Wooldridge's originality is the use he made of the effect of intravenous injection of the tissue-fibrinogen to explain the conditions seen in cardiac disease. Wooldridge had shown, as above stated, that the action of the tissue-fibrinogen upon the blood was to evoke two different phases of coagulation—namely, the positive phase of increase of coagulability, and the negative phase of diminution of coagulability; but he also recognised that the action of this complex albuminous substance was not merely an alteration in the fibrinogen formation, but that it also profoundly altered the blood, and in such a way as to change its behaviour towards the vessel-wall, or, to put the same thing into language more consonant with recent research on the formation of lymph (Heidenhain), changed the behaviour of the vessel-wall towards the circulating blood.

The alteration of the natural relation between the circulating fluid within the blood-vessels and the tissue juices which he discovered to be brought about by the intravenous injection of tissue-fibrinogen consisted in a severe transudation in any part in which the circulation was mechanically hindered on the venous

side.   Experimental research has long shown the absurdity of
the ordinarily received clinical notion that simple mechanical
interference with the venous outflow from any given area is
sufficient to cause œdema.   Although this has now been demonstrated for over twenty years, nevertheless that time has
not been sufficient to eradicate the falsity[1] of the supposed
principle.   The special pathological condition which is supposed
to illustrate it is essentially chronic cardiac disease, in which
there is considerable venous obstruction, and yet there is, of
course, no scientific evidence in favour of this position.   Some
factor has in such cases been overlooked, and, of course, unfortunately, clinical investigation in the way in which it is ordinarily carried out offers no solution of the problem.   The venous
stagnation is obvious; the question for answer is, What is
probably the missing factor?   Some valuable observations by
Wooldridge are extremely suggestive in this particular.   He
found that if tissue-fibrinogen were injected into the circulation
of an animal, and if the circulation through one of the limbs be
mechanically obstructed by ligature of the femoral vein, a most
extensive and rapidly developed œdema of the leg occurs, and
very often there is in addition exudation of red blood-corpuscles.
Such œdema and petechial hæmorrhage clearly results from the
direct toxic effect of the tissue-fibrinogen on the blood.   He
further discovered that this effect was developed proportionately
to the rapidity with which the tissue-fibrinogen was injected
into the circulation.   Of course, the experiments of Lichtheim
negative the notion that the rapid injection of 12 c.c. of the
solution can by virtue of the quantity of fluid produce the
slightest effect; the influence of rapidity therefore must be due
to the blood requiring to have a large quantity of the toxic agent
applied to it at any given moment, to set up the above disorder

[1] *Vide* the recent discussion on Dropsy at the Roy. Med. Chir. Society, May
1892.

of relations between the vessel-wall and the contained fluid. The actual proof of this is established by the fact that dilution [1] of the solution of tissue-fibrinogen prevents the resulting œdema in the same manner as does the slower injection. According to Wooldridge, it is possible that the tissue-fibrinogens contained in the lymph-glands, and under normal circumstances constantly being discharged into the circulation and therein disintegrated, as his experiments go to show, might, under the circumstances of a greatly disordered circulation by the mechanical obstruction of a valvular disease, exert an unfavourable, i.e. toxic, influence. His experimental observations naturally show a good ground for such a very original idea. It is greatly to be hoped that this, which was almost the first suggestion towards a rational and intelligent pathology of this common and severe disease state, may be followed up by other research, which cannot fail to have a fruitful outcome.

[1] The importance of concentration of the fibrinogen solution as a factor in its influence on the blood has been confirmed by Wright (*Proc. Roy. Irish Acad.* 1892, p. 139).

Part I

PHYSIOLOGICAL PAPERS

PREFATORY CLASS

ON THE FUNCTIONS OF THE VENTRICULAR NERVES OF THE MAMMALIAN HEART

ON THE FUNCTIONS OF THE VENTRICULAR NERVES OF THE MAMMALIAN HEART [1]

In the descriptions of the cardiac nerves given in anatomical text-books, mention is made of branches which pass across the auricle on to the ventricle.[2] So far as I am aware, the functions of these nerves have not hitherto been discussed in physiological literature, and this is certainly partly due to the fact that, in spite of their size, they are invisible until special methods are employed. After Professor Ludwig had shown me a means of rendering them visible in any animal that had been recently killed, and had thus enabled me to become acquainted with their origin and distribution, I determined to investigate their physiological functions. Before recording the results of this research, I will relate the issue of the anatomical observations, which were carried out on dogs.

The nerves which go to the heart and to the large vessels may unite in their course with others of the same side to form a plexus, or they may also join those of the other side as well. According to the place where they terminate, they may be classified into nerves of the large vessels, auricular nerves, and nerves which continue on to the ventricles either before or after giving off branches to the auricles. In accordance with the object of this research, the anatomical remarks will be chiefly confined to the ventricular nerves.

The fact that the mammalian ventricles are surrounded with

---

[1] [Translated from *Du Bois' Archiv*, 1883, p. 522.]

[2] These accounts are mainly based on the observations of Lee (*Philosophical Transactions*, London, 1849) and Schlkarewski (*Göttinger Nachrichten*, 1872, p. 426). The former has given a careful description, accompanied by excellent illustrations, of the ventricular nerves in man, and, so far as my experience goes, this system is identical with that of dogs.

a network of nerves situated immediately beneath the pericardium can only be observed without artificial aids in the case of very lean hearts. Special methods must be employed before they can be seen in well-nourished animals, particularly in dogs. A simple and sure way consists in painting the surface of the heart of a dog just killed with carbolic acid which has been liquefied by heat, the heart having been previously washed free from blood with a 0·5 per cent. solution of common salt. By this means the nerves stand out as white threads against a brownish background. If the whitish colour produced by the carbolic acid be likewise imparted to the surrounding connective tissue, the nerves disappear again—a process which generally occurs a few minutes after the painting with carbolic acid. In order to obtain an exact representation of the course of the nerves, it is advisable first to make a drawing of the heart, and then to subject a small surface at a time to the action of the carbolic acid. In this way figs. I. and II. were successfully obtained, and their accuracy assured by frequent comparison with other hearts treated in a similar manner. The separate nerves, which are mostly very fine, all consist of non-medullated fibres ; they form numerous plexuses and run down obliquely from the base to the apex of the ventricle, crossing the superficial muscular layer. As already mentioned, the minute trunks lie immediately below the pericardium, although a few are sunk in the muscular tissue of the heart.

The trunks which divide to form the nervous covering of the ventricles cross the auriculo-ventricular groove at three different points. One (OO in fig. I.) runs along the left side by the pulmonary artery towards the anterior longitudinal fissure. The larger number of its branches supply the anterior surface of the left ventricle, but a few cross the longitudinal fissure and are distributed to the surface of the right ventricle. The nerve-trunk, which has this distribution, is composed of fibres which enter the cavity of the thorax from both the right and left sides. In its course over the anterior surface of the auricle it gives off small branches to the left auricle, as well as one which enters the ventricular septum. A second trunk, of similar origin to the one just described, emerges between the roots of the large arteries, therefore to the right of the pulmonary artery, and spreads exclusively over the right ventricle (O, figs. I., II., and III.).

The third trunk divides as it is passing over the auricle, so that a bundle of nerves is here formed. This immediately breaks up into a number of fibres which spread over the posterior surface of the left auricle (fig. II. 5). Numerous branches issue from this network, a few of which run along the auriculo-ventricular groove; the greater number, however, are distributed over the posterior surface of the two ventricular walls, forming numerous anastomoses in their course.

If the ventricular nerves be traced to their origins, the results are as follows :—

A. The two nerves which stretch across the anterior surface of the ventricles (fig. I., O and OO) spring for the most part from a network lying behind the arch of the aorta. The plexus itself is formed from three sources :—

1. A large branch on the right side which issues either from the trunk or from the recurrent branch of the right vagus nerve, and sometimes from both at once (fig. III. 4', and fig. V. 4). These nerves always send some branches to the auricle.

2. One or two branches arising from the ganglion of the left vagus (fig. III. 3, and fig. IV. 3).

3. A short branch from the recurrent nerve of the left vagus (fig. III. 6).

Besides the branches (fig. III., 1 to 3) which proceed towards the plexus from which the nerves start to pass along the right and left sides of the pulmonary artery to the anterior surface of the ventricles, we frequently find

4. A branch from the recurrent nerve of the left vagus running alongside of the pulmonary artery to the larger nerve-trunk (fig. III., OO) which usually spreads over the anterior surface of the left ventricle.

B. The bundle of nerves (fig. IV. 5) which courses over the left auricle to spread out on the posterior surface of the ventricular wall is derived exclusively from the nerve-trunks on the left side of the chest. It arises either from the vagus ganglion or from the ansa Vieussenii, and sometimes from the first thoracic ganglion of the sympathetic. Before reaching the heart the nerve often sends branches to the auricle; this and other variations occurred in the heart from which fig. IV. was drawn. It also frequently appeared to me that the bundles running over the left auricle sent some branches to supply the auricular wall.

The description I have given of the origin and course of the nerves before they reach the ventricles is founded on numerous and careful dissections of dogs.

In order to ascertain the action exerted by the ventricular nerves on the heart, and from it reflexly on other organs, it was necessary to adopt suitable methods of stimulation. For this purpose only small or moderate-sized lean dogs can be used, on account of the numerous preparations required for the experiment. It is also desirable that the animal should be kept quiet, and therefore as free from pain as possible, although not so as to enfeeble the reflex action in response to a decided stimulus; and, finally, it is essential that the blood-pressure should be about normal. The dog was therefore narcotised by an injection of tincture of opium ; an incision was then made at the posterior end of one parietal bone close to the middle line, but avoiding the longitudinal sinus, and through this the brain was cut across behind the corpora quadragemina. This simple trephining operation, which is usually accompanied with very little bleeding, leaves the medulla oblongata with an increased reflex activity, and secures the normal tone of the blood-vessels. After these preliminaries tracheotomy was performed and artificial respiration carried on, and the heart with its nerves was exposed. For this purpose an aperture extending from the second to the fifth rib was sufficient, hæmorrhage being checked by ligatures in the following way :—

Before excising a piece of rib about as long as the finger, a fine brass wire or a strong thread was passed round each of the four ribs close to the sternum and at a little distance from the spinal column, the loops thus formed being drawn tight so as to completely close the intercostal arteries and thus prevent any hæmorrhage from them. The nerves to be stimulated or cut through were now dissected out in the pleural cavity by means of blunt needles or forceps, and a mark was attached to each in order to be able to trace its course by dissection after the animal's death. It was usually found necessary, even while the experiment was proceeding, to dissect out the ventricular nerves just before they reached the ventricular wall. This cannot be successfully accomplished unless the minute trunks be sought, isolated, and divided at the roots of the large arteries and on the posterior surface of the left auricular wall. When we come to cut through the nerves which issue near the pulmonary artery, we find them covered by the left auricle, which should therefore be drawn aside by an assistant by means of a thread tied to the extreme apex. In the space thus obtained, carefully avoiding all bleeding from the neighbouring coronary artery, the larger plexus lying on the left side of the pulmonary artery may now be found immediately above the auriculo-ventricular groove, and divided. The small branch which emerges at the right side of the pulmonary artery is best sought in the connective tissue between the aorta and the pulmonary artery, the search being materially aided if the latter be drawn a little to one side by an assistant, but, of course, not so much as would impede the flow of the blood-current from the heart. It is a far more arduous and difficult matter to find and divide the bundle of nerves on the posterior surface of the heart. The apex of the heart must be raised before the place where they pass on to the ventricles can even be seen, and while the heart is being adjusted in a suitable position, it may easily happen that the blood inside the large arteries is perceptibly checked and the cardiac beats thus rendered irregular. Should the heart show signs of stopping altogether, the operation must be discontinued until the ventricles, restored to their natural position, regain their power of beating. The greatest caution must be exercised in detaching the delicate nerve-branches from the layer of pericardium, in order to avoid tearing the very thin auricular wall. It is scarcely necessary to remark that the success of the

E

operation must again be tested after death, with the help of the carbolic treatment. In view of the difficulties of the operation thus described, it may be asked why the method of cutting through the nerves was chosen rather than the apparently simpler mode of destruction by means of corrosive substances or the galvanic cautery. These means were tried, but the uncertainty of the result with only a feeble application, and the disturbances caused to the cardiac circulation by a more powerful use, decided me to abandon the indirect method.

There is yet another operation, which consists in leaving the superficial ventricular nerves intact and in completely destroying all the others, so that the stimulation of the trunks of the vagus and accelerator nerves can only reach the ventricles by this one path. I attained a fair measure of success by the following means. A sound with a strong silken thread attached was inserted between the anterior surface of the auricles and the posterior surface of the large arteries. One end of the thread was taken round the apex of the heart to the posterior surface of the auricles, and the two ends were then drawn tightly together through a button. By this means the mass of auricular muscle may often be completely crushed without tearing the pericardium. As the anterior ventricular nerves run in front of the large arteries, they remain outside the ligature, and therefore uninjured. The ligature adopted by Stannius for frogs' hearts may thus be employed for the mammalian heart. But its use is not always attended with success; the thread may be drawn either not tight enough or too tight in the clamp, and thus either undestroyed bridges of muscular wall may be left or the wall may in some places be torn right across, so that the blood gushes out. Complete crushing of the muscular tissue without tearing the pericardium succeeds better in the rabbit than in the dog. It may also happen that the thread is tied round the transverse furrow, thus including branches of the coronary arteries, although this may be avoided by the exercise of care. The auricle cannot remain long confined without prejudice to the irritability of the ventricles; but if the preliminary measures have been attended with success, the stimulation of the vagus or of the accelerator nerve may be accomplished before the thread is removed. As a rule, these nerves were not stimulated until the loop of thread had been slackened.

I am still engaged in perfecting this experiment, since it gives promise of interesting results.

In all the experiments the cardiac beat and the arterial pressure were registered by a mercurial manometer attached to the carotid artery. A few remarks concerning the use of the manometer in recording cardiac movements may not be out of place here. The work performed by an isolated frog's heart can be estimated more completely by this method than by any other, provided that the ventricular contents at every contraction are only allowed to flow into the tube of the manometer. Under these conditions the product of the weight of the issuing fluid into half the height of the pressure attained represents the work done by the total mass of ventricular muscle. If, moreover, before each fresh systole occurs, the quiescent ventricle be supplied with a solution driven through the veins at a constant pressure, we are able to compare the work executed by different systoles, and can thus estimate the variations caused by changes in the muscular tissue or in the nerves by which it is excited. As it is naturally impossible to obtain these conditions in the case of the heart of living mammals, the manometer can here only be utilised to record the pulsations, for its other duty—that of showing the arterial blood-pressure—is known to depend upon various other circumstances besides the work done by the heart. To ascertain whether a change occurring in the arterial pressure during the course of the experiment is occasioned by the heart or by other causes, special measurements must be instituted or a variety of conditions upon which the state of the blood-pressure depends must be eliminated.

The nerves which proceed to the ventricles might act either in a centripetal or a centrifugal direction, and to decide this, stimuli of various kinds must be applied at different spots.

I will commence by describing the results of stimulating the peripheral stump of the nerves after they had been divided. In this series of experiments the nerves were stimulated by means of the induced current, and hence the necessity arose for exciting them in such a way as to protect the ventricular substance from the current. As it was accordingly impracticable to apply the induction-current to the nerves in their normal position, and as a diminution of their irritability was to be feared if the branches were picked up and drawn out, the trunks lying further away from the heart were stimulated in their stead. But as the latter send branches to both auricles and ventricles, measures were adopted to confine the action of the stimulated nerves either to the auricles only or to the ventricles only.

I. The trunks of the vagus or of the accelerator nerves were first stimulated while the cardiac nerves were untouched, and then again after the ventricular nerves on the posterior or anterior or on both surfaces of the ventricular wall, that I have described, had been completely cut through. Among a not inconsiderable number of experiments in which it was endeavoured to carry out this plan, of course only those were counted successful where the nerves were found to be all divided on careful dissection of the heart at the end of the experiment. The results were as follows:—

STIMULATION OF THE ACCELERATOR NERVE BEFORE AND AFTER DIVISION OF THE VENTRICULAR NERVES

EXPERIMENT 1.—*Stimulation of the right accelerator nerve at the point where it leaves the ganglion stellatum*

	Number of pulsations in every 5 seconds			Duration of stimu-lation.	Distance of secondary from primary coil.
	Before	During	After stimulation.		
Before dividing the ventricular nerves	14	14·5–16	19–18–17	12 sec.	6 cm.
After dividing the ventricular nerves	13·5	14–17	16·5	9	6
	13–13·5	13–15	17–16	12	4

EXPERIMENT 2.—*Stimulation of the right accelerator nerve near its exit from the ganglion stellatum*

	Number of pulsations in every 5 seconds			Duration of stimulation.	Distance of secondary from primary coil.
	Before	During	After stimulation.		
Before dividing the ventricular nerves	9·5	12–14·5	13	10 sec.	6 cm.
	11	14–14–12·5	12·5	13	6
After dividing the ventricular nerves	10	11–11·5	12·5	12	
	10	12–13–15	14	16	3

STIMULATION OF THE VAGI IN THE NECK BEFORE AND AFTER DIVISION OF THE VENTRICULAR NERVES.

In the case of the two animals upon which the experiments above described were made, the stimulation was applied to one of the vagi. The effects of the stimulus before and after cutting through the ventricular nerves were approximately the same. For instance, after the ventricular nerves had been divided, in one case the heart stood still for the whole period of stimulation (nine seconds), and in the other case during a period of seven seconds' stimulation.

STIMULATION OF THE RECURRENT LARYNGEAL BEFORE AND AFTER DIVISION OF THE ANTERIOR VENTRICULAR NERVES [1]

EXPERIMENT 3

		Before	During	After stimulation.	Distance of primary from secondary coil.
Before dividing the ventricular nerves	*a.*	13–13	5	15–15–15·5	25 cm.
	b.	11·5	2·5	10–15·5–15·5	24
After dividing the ventricular nerves		12	3	15–16–15·5	24
After dividing a few auricular branches		12	8·5–9·5	11–12·5	—

EXPERIMENT 4

	Before	During	After stimulation.	Distance of primary from secondary coil.
Before dividing the ventricular nerves	14	0	16·5–16·5–18–17	25 cm.
After dividing the ventricular nerves	14–14–14	0	15–15–15	25

[1] Its branches only go to the anterior ventricular nerves, therefore the division of the nerves on either side of the pulmonary artery is sufficient.

After the ventricular nerves had been removed the heart-beats were feebler and the pulse weaker.

As under normal anatomical and physiological conditions of the heart's structure and irritability, the beat of the ventricle is preceded and determined by that of the auricle, and as therefore the stimuli producing the pulsations originate in the region of the auricles in the case of the mammalian heart as well as in that of the frog's heart, there is no occasion to suppose the intervention of any nervous communication between auricles and ventricles in order to produce the standstill of the ventricles. If no stimulus arises in the auricle it cannot be propagated to the ventricle; both divisions of the heart will remain quiescent, and the division of any nerves we might choose could exert no influence on the results of stimulating the vagus.

The method of procedure I adopted suggested an experiment which appeared likely to assist in solving the question just started. It is well known that very delicate nerves go to the auricle. The small number of fibres of which each is composed, together with the simultaneous presence of a number of nerves all much alike, render it indubitable that each can only supply a very limited region, which, so far as it can be traced, lies in the auricle; but, unfortunately, owing to the inadequacy of our anatomical methods, no precise proof can be adduced, and it is possible that one or two of the branches may be continued on to the ventricle. At all events, it is more natural to assume that they are entirely restricted to the auricle than that they are thus distributed. We will now give the results of stimulating one of these small nerves.

STIMULATION OF SMALL BRANCHES OF THE VAGUS GOING TO THE AURICLE

EXPERIMENT A.—*The branches proceed from the trunk of the vagus nerve to the left auricle, leaving the trunk near the origin of the recurrent nerve*

Before	During	After stimulation.	Distance of primary from secondary coil.
13	9	13	25 cm.

EXPERIMENT B.—*Branches from the left vagus to the left auricle*

Before	During	After stimulation.	Duration of stimulation.	Distance of primary from secondary coil.
11	0	11	3 sec.	25 cm.

EXPERIMENT C.—*Branches from the right vagus nerve to the right auricle near the mouth of the superior vena cava*

Before	During	After stimulation.	Distance of primary from secondary coil.
16	5	16	25 cm.

Should it be definitely proved that the minute branches terminate in the auricle, the results of stimulating them would show that there was no occasion for any immediate action on any part of the ventricle in order to produce inhibition of its contractions. The inhibition of the auricles would at once cause that of the ventricles.

While this conclusion is still open to question, what I am about to describe may be considered to be definitely ascertained. For, as the stimulated nerve divides into branches to supply only a small section of the auricles, and as it nevertheless stops or slows the beats of the whole auricle, it is evident that the nerve must act on a spot from which the rhythm of both auricles is controlled. If it be permissible to have recourse to analogy for the explanation of this phenomenon, we may compare it to the reflex actions starting from the central nervous system. It is known that the contraction beginning at the auricle and proceeding to the ventricle, and generated by a stimulus confined to one spot of the ventricular surface, has often been regarded as a reflex action. In accordance with this view, the phenomenon observed by me should be added to the class of reflex inhibitions.

After this digression I will return to the experiments in which the nerve-trunks of the heart were stimulated before and after the anterior and posterior ventricular nerves had been cut through.

Even if the standstill of the ventricle following on that of the auricle can be explained without having recourse to a nervous communication, yet it cannot be doubted that the two parts of the heart, so far as regards its contractions, are connected by means of nerves. The quickening of the cardiac beat by the stimulation of the accelerator nerve may be brought about if this nerve is in direct connection with both auricle and ventricle at the same time ; otherwise, if we suppose that the nerve goes to the auricle only, there must be some means of communication between the two by which the impulse is propagated from above downwards. The nerves I have described cannot be reckoned among those which either directly or indirectly convey the impulses of the accelerator nerve to the muscular fibres of the ventricles, for stimulation of the latter nerve was as effectual after the ventricular nerves had been divided as before.

II.   The second series of experiments was a counterpart of the first. Of all the nerves going to the heart, only the branches proceeding direct to the ventricles were left connected with the vagi and accelerator nerves; the others were all divided between the place of exit from the trunks and of their supposed passage to the ventricles. An impulse produced in the vagi and accelerator nerves could therefore only be conveyed to the ventricular muscle by the branches which I have designated as ventricular nerves. Owing to the course taken by the anterior ventricular nerves, it is possible to perform on the mammalian heart the experiment of Stannius on the frog's heart without separating all the ventricular nerves from their trunks. For the method of ligaturing the auricle so as to crush its muscular wall and septum, and destroy its physiological continuity, I must refer the reader to page 50. Only the two anterior ventricular nerves were, however, left intact by the operation, but these were the ones that run to both right and left ventricle, and, unlike the third nerve, which was crushed, receive fibres from nerve-trunks of both sides of the body.

With the result produced by ligaturing the frog's heart at the auriculo-ventricular groove in my mind, I expected that crushing of the whole ring of auricular wall just above the valves would be followed by arrest of the contractions of the ventricles. The supposition proved erroneous. The ventricles as well as the auricles went on beating, but each at a different rate. My method of registering the heart-beats was inadequate to show the pulsations of both divisons, for the manometer inserted in the artery could give no account of the auricular pulsations, and it was also useless for the ventricle so long as the thread was drawn tight round the auricles. I shall not be able to give a complete account of the phenomena presented by the auricular and ventricular contractions until an apparatus can be adopted which will record simultaneously the pulsations of the auricle and also of the empty ventricle. At present I must be satisfied with counting the auricular beats and with taking tracings of the arterial pulse after the ligature round the auricle is relaxed.

As the ventricles after separation from the auricles still beat regularly and forcibly enough to maintain a moderately powerful blood-current, we were now in a position to observe whether stimulation of the vagi or accelerator nerves has any influence through the anterior ventricular nerves on the beat of the ventricles. These results may be seen in the figures.

LIGATURE AND COMPRESSION OF THE AURICLES IN THE AURICULO-VENTRICULAR GROOVE, LEAVING THE ANTERIOR VENTRICULAR NERVES INTACT

STIMULATION OF THE VAGUS NERVE

Dog

A and *B.* The auricles were crushed by means of a ligature which could be tightened by twisting it round a rod.

Auricles and ventricles went on beating, but at different rates. Stimulation of the vagus brought about auricular inhibition, but the ventricular beat was unaffected.

The ligature was removed; at first auricles and ventricles beat with a different rhythm, but they soon resumed their normal character, and stimulation of the vagus now occasioned the standstill of auricle and ventricle.

Dissection of the heart after death showed that in neither case had the auricular wall been destroyed by the ligature. This explains why the impulse did not pass on from the auricle while it was tied up, but did so on release of the ligature. The experience gained in these and other experiments of a similar nature led me to adopt a clamp with a screw action instead of merely tightening the ligature by hand.

E. Clamping of the auricles followed by relaxation of the ligature.

a. Before vagus stimulation the ventricle beats 22 times in 20 seconds.
   During   „   „   „   „   25   „   „
b. Before   „   „   „   „   23   „   „
   During   „   „   „   „   16   „   „
c.   „   „   „   auricle stands perfectly still for 20 seconds.
d. During stimulation the ventricles cannot be arrested even by the application of a more powerful stimulus to the vagus, but it was uncertain whether the beats were retarded or not.

EXPERIMENT F.—The ligature was put on considerably above the auriculo-ventricular groove. After it was loosened

a. The ventricles beat 20 times before vagus stimulation in 20 seconds.
   „   „   19   „   during   „   „   „   „
   The auricles   „   12   „   before   „   „   „   „
   „   „   0   „   during   „   „   „   „
b. The ventricles   „   15   „   before   „   „   „   „
   „   „   15   „   during   „   „   „   „
   The auricles   „   23   „   before   „   „   „   „
   „   „   0   „   during   „   „   „   „

RABBIT

C. The ventricles beat   10   times before vagus stimulation in 10 seconds.
   „   „   9–10   „   during   „   „   „   „
   The auricles   „   15–16   „   before   „   „   „   „
   „   „   0   „   during   „   „   „   „

D. The auricles beat more frequently than the ventricles before stimulation.
   The ventricles beat   7   times before vagus stimulation in 8 seconds.
   „   „   7·5–7·5   „   during   „   „   „   „
   The auricles   „   ?   „   before   „   „   „   „
   „   „   0   „   during   „   „   „   „
(The stimulation lasted 16 seconds.)

STIMULATION OF THE ACCELERATOR NERVE

DOG

Before clamping the auricles, the ventricles beat in every 5 seconds

Before	During	After stimulation.
9·5	11–11–12–11–12·5–13	14–13 times.

After compression and removal of the ligature the ventricles beat in every 5 seconds

Before	During	After stimulation.
6	6–5·5–5·5–6	5 times.

These experiments all tend to show that stimulation of the vagi and accelerator nerves has no effect upon the ventricular systole when their trunks are connected with the ventricles by means of the anterior ventricular nerves only. As after exclusion of the latter the ventricular beat is still affected by the vagi and accelerator nerves, but not when the impulse can only be conveyed to the ventricle by means of these anterior ventricular nerves, it follows that they have no power of altering the rhythm of the ventricular systole.

These results bring this part of my research to a close. It will be obvious that there are numerous questions and observations connected with the functional separation of the mammalian auricles and ventricles by means of a clamp, and these I shall hope to consider and recount in a further communication.

III. The third series of experiments refers exclusively to the posterior ventricular nerves. Although the results obtained with the anterior ventricular nerves led me to suppose that the third ventricular nerve would exert no influence on the rhythm and force of the systole, yet they did not furnish conclusive evidence on this point. The means adopted for the anterior ventricular nerves cannot be applied to the peripheral stump of the posterior one, because, being so closely connected with the wall of the left auricle, it is impossible to avoid crushing it at the same time. No other course remained, therefore, but to stimulate the peripheral stump of the nerve after it had been cut across, without further dissection from the auricular wall. As the nerve can always be found, the experiment was practicable, and I did not think it would be fruitless for the reason that it varies in its origin, sometimes deriving most of its fibres from the vagus and at other times from the accelerator nerve. Hence I hoped that out of a number of observations the nerve might sometimes be obtained and stimulated, perfectly free from accelerator or inhibitory fibres. This anticipation was justified by the result.

Of fifteen animals in which the peripheral stump of the posterior ventricular nerve was tetanised at some distance from the heart, stimulation produced—

Slowing of the pulse in three cases.

Acceleration in two cases.

No change in ten cases.

As the nerve has a varying origin, we may assume that in the first three cases vagus fibres, and in the next two cases accelerator fibres, had been mixed with it, but that in the larger number of cases this had not occurred, and the nerve had remained free from either.

If this deduction be accepted, it would appear that the posterior ventricular nerve has, like the anterior ones, no effect upon the rhythm of the systole. The conclusion would, however, only hold good if it were proved that the nerve maintained its normal irritability after stimulation with a negative result. With regard to this point, the question arises as to whether the irritability of the cardiac nerves generally is at all diminished by the operation. This may be ascertained by stimulating another nerve lying close to the posterior ventricular nerve, stimulation of which will produce definite results. The event proved that this fear was unfounded, for with the same animals in which stimulation of the posterior ventricular nerves had not affected the beats at all, slowing was immediately produced when the stimulus was applied to the branches of the vagus. The probability that the nerve retained its irritability was turned to certainty by the results of two out of the ten observations. In these stimulation again brought about no change in the ventricular rhythm, but caused a rise in blood-pressure. This seems to be conclusive evidence that the posterior as well as the anterior ventricular nerves exercised no influence on the rhythm of the cardiac beat.

The fact that in the two last-mentioned experiments, unlike the others, stimulation was followed by a positive result, is worthy of attention. Before considering the matter it will be advisable to give a table of the figures :—

STIMULATION OF THE PERIPHERAL STUMP OF THE POSTERIOR VENTRICULAR NERVE

Average of blood-pressure in every two seconds.

EXPERIMENT A

Before mm. Hg.	During mm. Hg.	After stimulation. mm. Hg.	Distance of secondary from primary coil. cm.
a. 50	64–78–30–82	82	6
b. 62	64–90–94–98–100	96–92	—

Pulse unaltered.

EXPERIMENT B

Before mm. Hg.	During mm. Hg.	After stimulation. mm. Hg.	Distance of secondary from primary coil. cm.
108–108 { 114–106–98–92– 88–88–88–86–86 }		82–84	6

| No. of pulsations in every two seconds | 5–6 | 5–5–5–6–4–4–4–4–4 | 4–4 | — |

It is evident from these observations, and particularly from the first one, that the increase in blood-pressure is due to the stimulation of the posterior ventricular nerve. But if this be so, it is curious that it should happen in such a small percentage of cases. It is possible that the nerve which causes a rise of blood-pressure may usually reach the ventricle by another path, and that it only occasionally runs in the posterior ventricular nerve. In support of such an assumption we might cite the statement of Pawlow,[1] who remarked that the stimulation of a cardiac nerve in the thoracic cavity was invariably followed by a rise of arterial pressure. His anatomical description, however, it seems to me, precludes the idea of the posterior ventricular nerve being responsible for the rise of blood-pressure.

In order to explain how it is that stimulation of the peripheral stump of a cardiac nerve can raise the blood-pressure, we must inquire whether the flow of blood into the vessels is augmented or whether the outflow from them is diminished. Unless we admit the not very probable theory that recurrent fibres of sensitive nerves run in the stimulated trunk, the only remaining alternative is to assume that stimulation has increased the amount of blood thrown out by the heart. This could be brought about either by a more abundant flow to the heart or by the ventricle emptying itself more completely in consequence of more forcible contractions. Although want of evidence prevents our deciding in favour of either explanation at present, we may call attention to the fact that the more powerful muscular contractions cannot increase the flow in the arteries unless the heart is at the same time supplied with a greater quantity of blood, as would happen, for instance, if the resistance in the pulmonary circulation were diminished.

---

[1] *Medicinisches Centralblatt*, 1883, p. 66.

But if the view be adopted that stimulation of the nerve has augmented the muscular energy of the heart, we should still have to decide whether the nerve acts directly upon the muscles or indirectly, by dilating the blood-vessels, and thus accelerating the blood-current through the coronary arteries, a possibility which has already been suggested by Pawlow. This idea induced me to stimulate repeatedly the peripheral stump of a cut nerve, which, so far as can be made out on dissection, is distributed to the cardiac end of the aorta. (*Vide* fig. IV. 4.)

The result of the experiment was negative in the case of three animals; blood-pressure and pulse remained the same even when a powerful induced current was employed.

STIMULATION OF THE CENTRAL STUMP OF THE AORTIC NERVE

	Before	During	After stimulation.
a. Number of beats in 5 seconds	11–12	8–9–7	7–11–13–15–13·5
Mean art. pressure in mm. Hg.	109	102–100	79–70–76–96–124
b. Number of beats in 5 seconds	14	9–9	8–10–13·5–14–16–16
Mean art. pressure in mm. Hg.	116	107	93–74–74–124

This negative result cannot be referred to want of irritability, since stimulation of the central end lowered the rate of the pulse and the blood-pressure.

After these negative results of stimulating the peripheral stumps of the ventricular nerves, they can no longer be regarded as motor nerves in the wide sense of the term. They must be classed among those cardiac nerves which are concerned in the production of sensation and reflex action, and this conclusion seems the more justified by their superficial course immediately beneath the pericardium, and especially by the fact that a sudden quivering of numerous muscles in the body takes place when the minute trunks are torn across. To the objection that there are a greater number of nerve-trunks than would be required for the slight sensibility possessed by the heart, we would reply by pointing out how intimately the heart, by means of the medulla, must be connected with the other organs of the animal body if it is to carry on its work without all manner of disturbances. To avoid these, the cardiac nerves, which give rise to reflex action, must be able at least to alter the rate of the beat, to regulate the flow in the coronary arteries, and to adapt the supply of blood flowing to and from the cardiac cavities to the irritability of the heart's muscles. A great number of nerves would be required by the heart to fulfil these numerous and, to some extent, contradictory requirements.

In order to gain an insight into the fibres in a nerve-trunk, excitation of which produces reflex action, the much-used method of L. Traube was adopted, the central stumps of the cut nerves being fastened between the electrodes of the induction apparatus and stimulated. As, according to my theory, the centripetal cardiac nerves are possessed of reflex-producing fibres of very varying action, it depends upon unavoidable peculiarities of the state of irritability as to whether the fractional portion of the possible reflexes which are notified by the manometer come into play at all, the greater irritability of one kind of fibre perhaps concealing that of another acting in a different sense, in the same way as observed with the centripetal pulmonary nerves. Even if this is not the case, our information must remain incomplete, since a whole series of phenomena exists which cannot be apprehended by the manometer. The results I attained exceeded my anticipations, for stimulation of the central stumps of different branches of the anterior ventricular nerves occasioned—

1. Slowing of the beat with rise of arterial pressure.

2. Slowing of the beat, the blood-pressure remaining unchanged so long as the stimulation lasted, the after-effects being an increase in the rate of the pulse and lowering of blood-pressure. The following figures may serve as examples:—

STIMULATION OF THE CENTRAL ENDS OF THE ANTERIOR
VENTRICULAR NERVES

1. *Branch of the right side.* (*Vide* fig. III. 4'.)

A.

	Before	During	After stimulation.
a. Number of beats in 5 seconds	17	15–14·5	15·5
Art. pressure in mm. Hg.	73	92–80	64
b. Number of beats in 5 seconds	17·5–18	15–14–14	14–16–16–17
Art. pressure in mm. Hg.	75–75	92–103–92	80–74

B.

Number of beats in 5 seconds	13–12	10–10–9	9–10–11–11–11·5
Mean art. pressure in mm. Hg.	76–73	72–73–71	57–54–46–48

2. *Branch of the left side.* (*Vide* fig. III. 3.)

Number of beats in 5 seconds	13	12–9–10	10–11–14–15
Mean art. pressure in mm. Hg.	66	62–58–52	48–50–58–80

Stimulation of the central stump of the posterior ventricular nerves produced—

1. Quickening of the beat without alteration of the arterial pressure.

2. Lowering of pulse-rate with diminution of pressure.

3. Slight rise of pressure without change in the beat.

4. Lowering of pressure without change in the beat.

STIMULATION OF THE CENTRAL STUMP OF THE POSTERIOR
VENTRICULAR NERVES. (*Vide* fig. V. 4.)

	Before	During	After stimulation.
a. Number of beats in 5 seconds	10–9–10	9–10–12–11–12	11–12–9–9
	No change in mean arterial pressure.		
b. Number of beats in 5 seconds	11·5–12	13–13–12–12	11
	No change in mean arterial pressure.		
c. Number of beats in 5 seconds	9–9	9–11–12	12–11–10·5–9·5
	No change in mean arterial pressure.		
d. Number of beats in 5 seconds	11–11	9–9–9–8	12
Mean art. pressure in mm. Hg.	64	62–48	72
e. Number of beats in 5 seconds	12·5–12	12·5–12–12	10·5–11·5
Mean art. pressure in mm. Hg.	96	102–102–98	92
f. Number of beats in 2 seconds	6·5–6·5–6	6–6–6–6·5–6–5·5–6–6	—
Mean art. pressure in mm. Hg.	92	98–88–80–72–68–60	—

The above observations may, I think, be taken to represent all the information which the manometer is capable of imparting, *i.e.*, rise and fall of tone in the branches of the aorta either alone or accompanied with acceleration, slowing or no change in numbers of pulsations.

Of a far more difficult nature is the solution of the other problem as to the place and conditions from and by which the different reflexes are started. We are not even possessed of the first requisite—the means of stimulation at any given and restricted spot without interference with the mechanism of the heart; nor have we so far any hope of attaining to such means.

Of the stimuli which have been successfully applied to the skin, only the slips of paper soaked in acetic acid have been found practicable for the cardiac surface. It is known that these produce contractions of the skeletal muscles, and even manifestations of pain, together with the usual consequences to the aortic current of stimulating sensitive nerves, viz. rise of blood-pressure.

In the meantime we must be content with the statement that stimulation of sensitive nerve-trunks proves that very various reflexes may be started from the heart.

PLATE 1.

To face page 67.

CLASS I

THE CHEMISTRY OF THE BLOOD-CORPUSCLES AND
THEIR RELATION TO THE PROCESS
OF COAGULATION

ON THE CHEMISTRY OF THE BLOOD-CORPUSCLES [1]

EACH of the three sections composing this communication might be read as a separate paper, yet they bear some connection with each other. In the first place, they were all undertaken at the instigation of Professor Ludwig ; and, in the second place, they all treat of the formed elements of the blood.

I. THE STROMA OF THE RED BLOOD-CORPUSCLES

As is well known, blood can be transformed in various ways into a 'laky' red fluid, e.g., by alternate freezing and warming to 55° C., by addition of a large amount of water or a little ether, by means of various acids, soaps, chloroform, or by the introduction of considerable quantities of urea, &c.

By these means the red corpuscles are split up into their two components, one of which is soluble in the blood serum, the other insoluble. The former consists, so far as we know, of hæmoglobin, and this is proved to be unchanged by the fact that it is still capable of altering its colour on the access of oxygen.

The other component, which Rollett designated ' the stroma,' still preserves the form of the original corpuscle, but has lost its colour and consistency, appearing as a delicate pale empty disc.

The statement we find in some text-books, that certain of the above-mentioned reagents also dissolve the stroma, is due to the extraordinary transparency of the 'laky' blood. We can show, however, that this is merely owing to the swollen

[1] [Translated from *Du Bois' Archiv*, 1881, p. 387.]

condition of the stroma, if we add a little free acid or
an acid salt. We then see a cloud form in the fluid, which on
microscopic examination is found to consist of the original
stromata. We can also easily recognise the presence of the
stromata if the ether which has caused their swelling up be
allowed to evaporate.

Since the stroma preserves its regular form under the
various conditions under which it is precipitated, we may regard
its composition also as uniform, and it would certainly have
been the subject of extensive investigation, considering the
great attention given by chemists to the red blood-corpuscles,
had it been possible to procure it in a pure condition and
in large quantities. This, I believe, I have been able to
accomplish.

Freshly defibrinated blood is mixed with many times its
volume of 2 per cent. salt solution and centrifugalised; the
precipitate of red corpuscles thus obtained is again washed
several times with salt solution on the centrifugal machine, so as
to get rid of the adhering traces of serum.

This mass, which consists of various forms of corpuscles,
is diluted with five or six times its bulk of water, shaken,
and treated carefully with ether till the fluid is perfectly
transparent. It is then centrifugalised once more in order to
separate the leucocytes, which swim in the fluid very little
altered. In order to be sure of their complete removal, the centrifugalising must be continued and repeated as long as any
turbidity appears at the bottom of the tube.

To the clear fluid thus obtained we add, drop by drop, a
1 per cent. solution of acid sodium sulphate. If the proper
amount of the salt is added, the clear fluid becomes as cloudy
and opaque as ordinary blood. Soon, however, the stromata
adhere together in clumps and sink to the bottom. Instead of
the acid salt we may use also dilute acids; the salt is, however, preferable, since with it there is less danger of decomposing
the hæmoglobin and thus contaminating the stromata with
hæmatin. After the precipitated stromata have shrunk, they no
longer swell up, even after long washing with distilled water, or

with water containing ether. As they can be easily filtered in this condition, we can extract all impurities from them, with the exception of a trace of hæmoglobin which has been precipitated with them.

All the operations here described can be carried out at a lower temperature in a few days, so we need not fear decomposition. At the same time it is advisable to purify the stroma and work with it as quickly as possible, since it is altered by the influence of distilled water.

The fresh stroma is entirely soluble in 0·2 per cent. hydrochloric acid; if it be kept some time under water in a cold place a part becomes insoluble in this acid. The part that is not dissolved by the hydrochloric acid resembles in its properties the nuclein-like body which we shall describe later.

The purified stroma consists of cholesterin, lecithin, paraglobulin, and a proteid compound with a body resembling nuclein; at times the stroma also contains traces of lime and iron (the latter probably due to the presence of hæmoglobin as an impurity).

1. *Cholesterin.* The filtered and pressed stroma is repeatedly extracted with cold ether. This æthereal extract on evaporation leaves needle-shaped branched crystals grouped in rosettes, which are transformed into the tablets characteristic of cholesterin when they are recrystallised from warm alcohol. These crystals show all the reactions of cholesterin, so that there can be no doubt about their nature.

The crystals might be contaminated with fats and with lecithin. The presence of the former was excluded by the absence of any smell of acrolein on heating the crystals in a glass tube with acid potassium sulphate, and also from the fact that the ether extract, evaporated to dryness, did not melt at 100° C.

Lecithin was proved to be absent by the fact that no phosphoric acid was formed when the ether extract of the stroma was ignited with sodium hydrate and saltpetre.

If traces of phosphoric acid can be detected by means of molybdic acid, the directions that have been given for the preparation of the stroma have not been strictly followed, and the leucocytes are not entirely got rid of.

The stroma can only be freed from cholesterin after repeated shaking with ether.

Petroleum ether may be used instead of sulphuric ether; by its means the cholesterin may be extracted from the stroma, free from fats or bodies containing phosphorus.

2. *Lecithin.* When ether will extract nothing more from the stroma, we may proceed to the extraction with alcohol; to carry this out fully, the stroma must undergo repeated and prolonged treatment with alcohol. Instead of extracting the stroma beforehand with ether, it may be treated at once with 80 to 90 per cent. alcohol at a temperature of 45° C. In this case the cholesterin separates in the crystalline form from the first portions of the extract on cooling. The filtered alcoholic extract is evaporated to dryness at 40° to 45° C. If the residue is now dissolved in absolute alcohol, any hæmatin that may be present remains undissolved. The fluid is filtered to separate the hæmatin, and the alcohol allowed to evaporate. We then get a yellowish waxy residue with the following properties. It burns with a bright flame, leaving an ash rich in phosphorus. In water it swells up, with the formation of the well-known myelin drops. It is more easily soluble in sulphuric and petroleum ether than in hot alcohol.

With platinum chloride its alcoholic solution gives a yellowish-white crystalline precipitate. This is shown to consist of lecithin platinum chloride by dissolving it in chloroform, then evaporating the chloroform and igniting the residue with soda and saltpetre, and determining the platinum, chlorine, and phosphoric acid.

The numbers were, Platinum	.	.	.	.	0·019
,, ,, Chlorine	.	.	.	.	0·022
,, ,, Phosphorus	.	.	.	.	0·007

These numbers show a proportion of 1 platinum, 6 chlorine, and 2 phosphorus, which is the proportion required for Strecker's formula for the double platinum salt of lecithin.

$$= 2(C_{44}H_{83}NPO_8Cl) + PtCl_4.$$

The fact that the precipitate produced by platinum chloride is almost entirely soluble in chloroform shows that the alcoholic extract of the stroma, after exhaustion with ether, contains little else but lecithin.

A body containing phosphorus has often before now been extracted from the red precipitate of corpuscles, and has been designated 'protagon,' but it has more recently been recognised as lecithin. Whether this belonged to the white or red corpuscles remained doubtful; and it is indeed improbable, having regard to the methods employed, that the lecithin was extracted from the red corpuscles, since these, as we have seen, give up their lecithin not to the ether but only to the alcohol.

3. *Paraglobulin.* If the stroma be treated with a NaCl solution after it has been freed by means of the water and ether from all foreign impurities, paraglobulin is taken up by the solution.

The quicker the purification has taken place, and the shorter the time that the stroma has been standing under water, the more easily can we remove all the paraglobulin.

The concentration of the NaCl solution is not a point of indifference. In my experience a solution of 5 per cent. is most useful, since in it the insoluble part of the stroma does not swell up so much, so that it can be easily separated by filtration from the fluid. The body that has been dissolved consists of paraglobulin. If the solution is saturated with NaCl, a precipitate is produced, and if this is filtered off, a second precipitate may be produced by saturation with magnesium sulphate; a precipitate is also produced if the 5 per cent. solution is diluted with water.

The precipitated substance shows all the reactions of an uncoagulated proteid. Dissolved in a 5 per cent. NaCl solution, it begins to coagulate at 66° C., and the coagulation is complete at from 69° to 70° C.

My experiments thus confirm an observation of Hoppe-Seyler, who proved the presence of a globulin in the stroma.

4. When the fresh stroma has been treated with 5 per cent. salt solution till nothing more can be extracted, a considerable amount is left undissolved. This residue is easily soluble in

0·2 per cent. HCl and in dilute alkalies. If we add to this HCl solution a proper quantity of glycerine pepsin, and then warm the whole for several hours to 40° C., a cloudy precipitate is produced. This is separated by filtration, and the filtrate is found to contain peptones. The small amount of residue on the filter contains sulphur and phosphorus, and is easily soluble in dilute alkalies, provided that it has not been previously treated with alcohol.

It is insoluble in artificial gastric juice, however long this is allowed to act on it at the temperature of the body. With nitric acid and ammonia it gives the xantho-proteic reaction. Thus the substance remaining after removal of paraglobulin, lecithin, and cholesterin, resembles in its behaviour the substance found by Plosz in liver-cells, and named by him ' nucleo-albu-min.'

Instead of that just described, we can also use a shorter method. Fresh stroma, washed with distilled water, is extracted with ether, and then dissolved in 0·2 per cent. HCl. The liquid thus obtained is generally somewhat cloudy, but if it is purified by filtration, it still retains its transparency even after standing for a long time.

If, however, it is digested for some hours with pepsin, a reddish cloudy precipitate is produced. In dilute alkalies or sodium carbonate this forms an opalescent solution which can be easily filtered.

Corresponding to its mode of preparation, this precipitate still contains lecithin, which can be extracted from it by alcohol, though not by ether. But the lecithin is extremely slowly extracted even by the alcohol, and the extraction is only complete after long-continued boiling.

The body freed from lecithin has lost its solubility in dilute alkalies, but otherwise it shows all the properties we have just detailed. It is worthy of mention that the same body may be prepared from the fresh stroma if this be dissolved in dilute alkalies before being exposed to the action of the artificial gastric juice.

The properties of the body left in the stroma after the

removal of the cholesterin, lecithin, and paraglobulin, lead us to conclude that it consists of a proteid which is combined with another molecule rich in phosphorus. Whether this latter is identical with the nuclein discovered by F. Miescher, I dare not decide, even though they have many reactions and properties in common. The body allied to nuclein is present in the stroma in such small quantities that, in spite of the large amounts of stroma that I worked with, I found a more accurate study of this substance impossible.

Although the knowledge we have gained of the constituents of the red corpuscles and their behaviour is not sufficient to afford us a perfect knowledge of the framework of the red blood-discs, yet it will give us more definite ideas to start from in future research on the subject. From this point of view the following remarks may deserve attention. Between the constituents of the blood-discs, which in laky blood go into solution, and those of the stroma which remain undissolved, there can exist no chemical combination even before the separation. At least it would be very hard to reconcile our ideas of chemical affinity with the fact that the blood-discs always split up into the same parts, however diverse the decomposing means may be, and especially whether this merely consists in a change of temperature, or in the presence of foreign molecules of very diverse composition and behaviour.

The decomposition is quite conceivable, however, if we consider that the soluble substances of the corpuscles are inclosed in an insoluble body, which hinders their passage into the solvent. We thus see that every time the insoluble substance changes its shape by swelling up, the soluble substance within may be pressed out, or, by the addition of the fluid imbibed by the corpuscle, may become more diffusible and so get out.

For a more accurate insight into the displacement of the contents than we have here assumed, a knowledge of the shape of the stroma in its swollen condition would be of great importance. Were it possible to make the stromata visible before their contraction, it would be at once possible to decide whether the hypothesis I have put forward could be upheld.

No less curious is the relation of the substances of the stroma to one another and to those which become soluble when the stroma is broken up. Why does not the paraglobulin leave the stroma together with the hæmoglobin, seeing that they are both equally soluble in serum, and the paraglobulin is dissolved easily enough in the NaCl solution, after the stroma has been treated with distilled water? The probable explanation is that the paraglobulin is present in the blood-discs in a compound which is split up by the distilled water. How and with what it is combined remains, however, uncertain. The lecithin, too, must be attached to the stroma, and especially to the nuclein body, in some peculiar fashion, as it would otherwise be inconceivable why it cannot be extracted by ether, and only with such difficulty by alcohol.

The proteid in the body which is left after the removal of lecithin, cholesterin, and paraglobulin must be in much more stable combination with the nucleinlike molecule, since the decomposition cannot be effected by mere solvents. This was only brought about, as we saw, when the proteids were converted into peptone by means of pepsin.

Thus the action of pepsin is not confined to converting insoluble into soluble proteids, but it is also able to separate a proteid from its combinations.

The act of dissolving one substance after another out of the blood-discs gives us no nearer knowledge of their anatomical structure. With whatever medium we treat the corpuscles or stroma, their appearance remains always the same, though, of course, they become more shadowy in proportion to their loss in weight by the solvent.

II. THE QUANTITATIVE ESTIMATION OF THE
COLOURLESS BLOOD-CORPUSCLES

IN order to determine the proportion of the weight of the dried constituents of the blood formed by the colourless corpuscles, I propose a method which is based on the employment of ether, a solution of magnesium sulphate, and the centrifugal machine. The ether, which destroys the coloured corpuscles, has no power on the leucocytes, so that when the blood has been rendered laky, these latter are the only structures kept in suspension by the fluid, and can be separated by means of the centrifuge.

Although I have worked thoroughly at the elaboration of the method of determination based on this behaviour of the white corpuscles, yet, owing to the short time at my disposal, I have not been able to satisfy all the requirements which are expected to be fulfilled by an analytical method; on the other hand, I think the observations contained in the following communication justify me in saying that the method pursued affords prospect of reliable results. After the colourless corpuscles have been separated from the blood by means of the centrifuge, they must be repeatedly washed to ensure their purity. We must first try to discover the changes they undergo under the influence of the fluids necessary for this purpose.

If dog's blood, prevented from clotting by injection of peptones, be centrifugalised, the colourless corpuscles collect together to form a disc at the junction of the plasma and the layer of red corpuscles (*cruor*), as Fano and Schmidt-Mülheim have already described.

Single cells, which may be artificially set free from the mass, possess at first living properties; if, however, they are allowed to remain for a day or longer in contact with the red corpuscles at the ordinary temperature of the room, they undergo a disintegration. The outset of this is marked by the area where the colourless cells come in contact with the red corpuscles assuming a venous hue, in contrast to the bright arterial colour of the rest of the mass.

Following this change of colour in their immediate neighbourhood, we find a change of the cells themselves, their bodies disintegrating to form a fibrinous clot, while their nuclei remain apparently unaltered.

If we add sufficient ether to peptone-blood to make it laky, the blood, as a rule, clots before the first stages of cleansing the corpuscles have been carried out. The necessary treatment of the cells is rendered practicable if we add to the blood a solution of $MgSO_4$ saturated at the temperature of the room and diluted with an equal bulk of water. We may call a solution of this strength 'half-saturated.'

If peptone-blood has been treated with a half-saturated solution of $MgSO_4$, its colour can be changed by means of ether without any clotting taking place. If it is then centrifugalised, a white disc floats on the surface of the red liquid, which, as the microscope shows, consists of a mixture of nuclei, granular masses, and a few fibrils. The granules and nuclei may be stained with hæmatoxylin.

In order to obtain leucocytes from normal blood, the latter must be prevented from clotting by admixture with a solution of neutral salt.

If the blood is allowed to flow from a vein into an equal volume of a half-saturated solution of $MgSO_4$, or a 10 per cent. solution of NaCl, it remains, as is well known, quite fluid. The effects of the two salts, however, are not quite similar, for clotting begins in the NaCl solution as soon as the mixture is diluted with three or four times its volume of water; while the blood that has been mixed with half-saturated $MgSO_4$ solution may be diluted with any quantity of water without clotting taking place.

If we assume, with Alexander Schmidt, that the presence of the ferment which is set free in the disintegration of cells is necessary for clotting to take place, we must conclude that, if they have once been in contact with a concentrated solution of $MgSO_4$, the leucocytes are protected from the deleterious influence of a more watery solution, which in the fresh state they are unable to withstand.

It can be also shown by a simple experiment that there is no ferment present in the blood treated with a half-saturated $MgSO_4$ solution.

If by means of the centrifuge all cells be separated from the mixture, and the fluid be diluted with eight times its volume of water, it remains perfectly fluid for days, but clots in a few minutes if a little ferment (prepared by Schmidt's method) be added to it. For my purposes, therefore, the best course was to use $MgSO_4$ exclusively.

The blood diluted with half-saturated $MgSO_4$ solution can be rendered laky with a much smaller quantity of ether than is required to effect this change in pure blood.

If the laky mixture be centrifugalised for several hours, the same colourless disc separates out as was the case with peptone-blood under similar conditions. The disc consists, as in the latter case, of nuclei, disintegrated protoplasm, and a few inconspicuous fibrils, which are fused together to form a coherent mass.

The evident changes which the leucocytes of normal and peptone-blood have undergone after the treatment adopted, are to be ascribed not to the ether but to the $MgSO_4$. The truth of this can be shown with the help of defibrinated (whipped) blood. Many leucocytes, as is well known, escape the destruction which accompanies the clotting of the blood ; these we find unchanged, heaped together at the bottom of the vessel, after the blood has been made laky by ether and centrifugalised for some hours.

To ensure the success of this experiment, not only must the blood be used as quickly as possible after it has left the veins, but it must not be left too long in the centrifuge.

I cannot help thinking that the accumulation of the cells in one spot caused by the action of the centrifuge, leads to a fusion of the individual cells; whether through the pressure excited by the centrifugal force, or by the mutual interaction of the cells where they touch one another, must remain a matter of speculation.

From the mass of cells which has been obtained with the help of the salt solution, the $MgSO_4$ must be removed. For this purpose I have used water containing some ether in solution, after assuring myself that this procedure caused no further alteration in the cells that had been treated with this salt. The advantage of this mode of treatment with $MgSO_4$ is well seen if we wash the colourless cells of whipped blood with ether-water without previously treating them with the salt solution. If the laky blood diluted with several times its bulk of water be centrifugalised, a white body is precipitated at the bottom of the glass cylinder, which has all the appearances of ordinary fibrin. The clot has become moulded to the shape of the vessel in which it was formed; if we throw it into water, it unravels to form delicate transparent pellicles, which under the microscope are seen to consist of a network of fine fibrils enclosing unchanged nuclei.

Thus every one of the fluids used to purify the white cells effected some alteration in their structure; and though the $MgSO_4$ solution caused the least amount of disintegration, yet its use arouses some feelings of doubt, for we could only be perfectly certain that the cells had suffered no material loss in the process of purification if they were still living at the completion of these processes. Since at present we cannot obtain this condition, we must seek in some other way to prove, or at any rate to show the probability, that the cells have neither lost nor gained in weight during this alteration of their structure.

My reasons for believing that the weight of the unchanged cells corresponds to that of the cells after their purification has been completed, are based on the following experiments.

Since it is at present still impossible to prepare perfectly pure leucocytes from unclotted blood, I was compelled to extend my observations to a closely allied variety of cells, those of the lymph-glands.

The lymph-glands of recently killed animals were washed with 5 per cent. NaCl solution, chopped up, put into linen bags, and the cells kneaded out through the linen. All that had been passed out was shaken up with 0·5 per cent. NaCl solution and centrifugalised. The precipitate found at the bottom of the cylinder, after it had been once more washed with 5 per cent. NaCl solution on the centrifuge, consisted almost entirely of lymph-cells. Their mode of preparation precludes any idea of their being still alive, nor can we determine what or how much has been given up by the cells to the NaCl solution used in the washing.

It is certain, however, that the cells have exactly the same microscopical appearance before and after this treatment; the essential part of their structure is thus not injured by the 0·5 per cent. NaCl solution. Now these cells are broken up by the action of salt solutions (especially half-saturated $MgSO_4$ solution) and by the plasma of peptone-blood in such a manner that, while their nuclei are preserved unchanged, their bodies are converted into a fibrinous clot. In this transformation they neither lose in weight nor do they take up anything that can be weighed from the fluid by which they are broken up; for the clot containing the nuclei in its meshes, when dried, weighs the same as the cells did before they were given over to destruction.

If now we compare the action of the $MgSO_4$ on these leucocytes with its action on those of the blood, we cannot fail to recognise the similarity of the two processes. There is only one difference between the changes undergone by the two kinds of cells, and that is in my favour. Apart from the fact that in both cases the nuclei are unaffected, the changes undergone by the colourless blood-corpuscles never go as far as the production of fibrin. Between the nuclei lies a mass whose microscopical appearance has the greatest similarity to protoplasm.

The mass of substance formed from the protoplasm of the cells far exceeds in amount that formed by the nuclei; and this can be made out with certainty, since everything that has been formed from the cell has been pressed together into a heap by the centrifuge. Indeed, from the microscopical appearance of small portions of the mass, it seems probable that the protoplasm has not, after the breaking up of the cell, left the nucleus at all. Never do we come across free nuclei; they are always embedded in the substance resembling protoplasm, and as a rule the individual nuclei are separated from one another by the distance of one diameter of a white corpuscle.

Although these experiments afford us no proof that the fluids used for washing have not removed any organic or mineral substance from the colourless blood-corpuscles, yet they leave no doubt that the mass remaining after complete purification of the cells contains the essential part of the fixed framework of the cells, especially the proteids, lecithin, cholesterin, nuclein, &c.

So long, however, as we have no proof that the colourless substance remaining after the washing process contains everything that was present in the uninjured cells, we can only allow a relative value to any method of estimation; for at present it can only rank as a process in which not the whole mass of the cells, but merely a fraction of their weight, is determined, though it is highly probable that this fraction bears a constant ratio to the whole amount.

Perhaps this probability might be raised to a certainty by determining the weight of nuclei in the mass remaining after the washing by Miescher's method. I have not yet, however, made any observations in this direction.

Having dealt with the most important arguments that can be urged against my methods, I will now proceed to the consideration of some other possible objections.

Perhaps the centrifuge may not have sufficiently separated all the insoluble constituents of the white corpuscles from the laky fluid. A doubt on this point can be easily removed by repeated employment of the centrifuge. If the laky blood remained in it the first time sufficiently long, we cannot obtain a trace of further precipitate from the clear fluid (decanted off from the precipitate), however many times, or however long, we place it on the centrifuge, or however long we leave it to stand in the ice-chest.

We might also suspect the presence of stromata as an impurity. It is easy to guard against this source of error by adding ether to the fluids used for washing; in this way the stromata swell up so much that they cannot be precipitated by the centrifugal force from the fluid. Since the stromata are soluble in 0·2 per cent. HCl, which does not dissolve the colourless cells, we have here also a method for assuring ourselves of the absence of stromata.

The ether, which must be added in order to get rid of the red corpuscles, might produce a precipitate in the plasma. Never, however, have I observed any clouding of the serum, salt- or peptone-plasma on adding 5 to 8 volumes of ether to 100 of these fluids, and none of the fluids used by me have contained a greater amount of ether than that. Nor is any precipitate produced in peptone-plasma or serum when they are mixed with an equal volume of a half-saturated solution of $MgSO_4$. The same may be said of salt-plasma, for this too becomes turbid only where the amount of magnesium sulphate it contains is considerably increased. Finally, as we have already mentioned, no precipitate is produced by the water used to dilute the magnesium sulphate plasma.

We may also mention the following characteristic properties of the colourless cells and of their *débris* that have been prepared in the manner described: they are insoluble in 0·2 per cent. HCl, and also in aqueous solutions of NaCl and $MgSO_4$. In artificial gastric juice, a part is dissolved after digestion, the rest, a not inconsiderable amount, being quite insoluble. The whole mass may be dissolved by dilute alkalies. Cold alcohol extracts from it lecithin, cholesterin, and probably a third body; when calcined, the mass leaves an ash containing lime.

After these general observations, I will now describe more exactly the procedure adopted in the analysis I shall afterwards give.

The colourless cells were estimated in unclotted, in whipped, and in peptone blood. If the blood is to be used before clotting, it must be received with especial precautions into the half-saturated $MgSO_4$ solution. The interval from the moment at which it leaves the carotid artery to that at which it reaches the salt solution must be as short as possible, and mixture of the two fluids must be as thorough and rapid as possible. It is therefore necessary to insert a short straight glass tube into the artery, to allow the first few drops of blood to escape freely, and then to receive the amount wanted for the determination directly into the graduated cylinder, carefully avoiding all frothing.

This procedure requires a powerful stream of blood; on this account, if large quantities of blood are required, we must always employ large dogs. If smaller animals are used, we may only obtain quantities of blood so small that their removal has no depressing effect on the circulation. Any stoppage to the flow is to be carefully avoided, since the results of all the following chemical operations are more successful the shorter the time taken by the blood in passing from the vessels to the salt solution.

The whipped blood, too, was treated with $MgSO_4$ solution before centrifugalising, in order to impart to the cells the properties dependent on the presence of this salt. If we add one volume of a saturated solution of $MgSO_4$ to three or four volumes of the blood, the white corpuscles rise to the surface of the fluid in the centrifuge, instead of sinking to the bottom, as would be the case without this admixture. This mode of separation takes place sooner, and the disc floating on the surface is more easily separable from the rest of the blood—another advantage of the use of $MgSO_4$ solution. The cells separated by the first centrifugalising are still further washed with the aid of the centrifuge.

At first I used for this purpose the $MgSO_4$ solution, for fear that clotting might occur in the remainder of the plasma still clinging to the colourless corpuscles. After it had been shown, however, that water containing ether would not precipitate any clot from the salt-plasma if this had been carefully prepared, I used this fluid for washing the cells, since the desired end is sooner attained by this means. The centrifugalising and the addition of water and ether were continually repeated till the precipitate was perfectly colourless, and the washings contained no traces of magnesia.

Now and then the fluid showed at this stage a slight cloudiness, which, however, must have been due to the slightest possible amount of solid matter, as many c.c. of the fluid on evaporation left the merest traces of solid residue, which in no case, as far as I could make out, contained any proteid.   Before the purified cells are washed out on to the weighed filter, they are treated with some alcohol, after which they can be quickly filtered without any loss.   The alcohol that runs through it is evaporated to dryness on one of the two watch-glasses, between which the filter containing the cells is to be placed for the operation of weighing.   The drying and weighing are carried out with the usual precautions.

I performed the three first estimations of the colourless corpuscles on blood that was prevented from clotting by receiving it into salt solution ; every two portions of 50 c.c. were received, one directly after the other, out of the same carotid. In every 100 c.c. they contained—

	A. grm.	B. grm.
I.	0·62	0·63
II.	0·57	0·59
III.	0·82	0·83

The agreement of the numbers encouraged me to compare the amounts of cells present in unclotted and in defibrinated blood.

The specimens of blood which were to serve for this comparative estimation were either taken simultaneously out of the two carotids, or else one immediately after the other out of the same carotid.

In 100 c.c. blood the amount of white corpuscles found was :—

		Unclotted blood. grm.	Clotted blood. grm.
I.		0·39	0·11
II.	a.	0·72	0·29
	b.	0·40	0·30
III.		0·40	0·29
IV.		0·54	0·39

All these numbers agree in proving that defibrinated blood contains a smaller weight of white corpuscles than blood which has not clotted, thus confirming the views of Alexander Schmidt, and at the same time giving them a quantitative expression. Yet I must mention the fact that in one of my estimations I found the difference in the amounts of cells in the well-clotted and in the unclotted blood much smaller than was found in the four preceding experiments.

I cannot ascribe this to an error in the analysis, since both specimens gave similar results:—

Unclotted blood. grm.	Clotted blood. grm.
V. *a.* 0·49	0·41
b. 0.48	0·45

In the first four experiments the difference in the weights of cells in the clotted and unclotted blood was sufficient to allow the supposition that the whole weight of the precipitated fibrin might be furnished by the cells that had disappeared. The fifth, however, no longer suggests this hypothesis. Are we to assume that fibrin may originate in different ways in different specimens of blood, or is it possible that a disintegration of cells can take place in blood treated with salt solution without a corresponding formation of fibrin? Future researches must decide this question.

It was proved by the observations of Schmidt-Mülheim and Fano, who injected peptone into the blood, that this proteid, when introduced into the blood, disappeared for the greater part, at any rate from the plasma. It also occurred to the latter of the two observers that the explanation of this fact might be that the peptone had been taken up by the colourless corpuscles. This hypothesis of Fano might be tested by determining the amount (in weight) of the colourless corpuscles of the blood before and after an injection of peptone. To carry out this experiment, however, two separate bleedings at different times would be necessary, since the normal blood, to serve for comparison with the peptone-blood, must be taken before the injection of peptone is carried out. If the experiment be carried out adroitly, however, very little time need intervene between the two bleedings.

Now I had already learnt from a part of the observations I have just described, that samples of blood obtained by bleeding at intervals of a few minutes contained equal amounts of leucocytes, so that the fact that the peptone blood was obtained a few minutes after the specimen of normal blood would make no difference in the comparison of their respective compositions.

The experiment was carried out as follows :—

The animal was first bled from the carotid into salt solution ; a 10 per cent. solution of pure peptone (prepared by Dr. Grübler) in 0·5 per cent. NaCl solution, in the proportion of ·3 grm. per kilo. body weight, was then injected into the jugular vein, in which a glass cannula had already been inserted.

Three minutes later a small amount of blood was allowed to escape from the carotid, and then the quantity required for analysis allowed to flow into magnesium sulphate solution in the same proportion as for the first specimen of blood.

All the other steps of the operation were the same as I have already described.

100 c.c. blood contained the following amount of cells :—

	Normal blood. grm.	Peptone blood. grm.
I. *a.*	0·46	0·59
b.	—	0·60
II.	0·39	0·57
III.	0·31	0·41

From the greater weight of colourless corpuscles in peptone-plasma, we may conclude that a portion of the peptone has been taken up by the colourless cells for the plasma. This would confirm not only Fano's hypothesis, but also a suggestion put forward by Hofmeister. The only way this observer can conceive of the passage of the peptone, formed as a result of proteid digestion, into the blood or chyle, is that it is taken up by the leucocytes.

Yet though my discovery may agree with these assumptions, it in no way affords a distinct proof of them. We can only consider them distinctly proved when we can again prepare the peptone out of the leucocytes.

At the same time it is apparent, from the comparison of the weights of the cells in normal and in peptone-plasma, that the difference is not sufficient to account for the disappearance of all the peptone that has been injected into the blood.

Whatever decision of this question the future may bring, further observations on the point as to whether peptone is taken up by the leucocytes is much to be desired, and would pave the way for a more definite knowledge of the changes undergone by this body in the organism.

III. *THE CONVERSION OF COLOURLESS CORPUSCLES INTO FIBRIN*

WE know from the famous researches of Alexander Schmidt on coagulation that fibrin and a similar substance may be produced in the disintegration of leucocytes. F. Meischer, too, in an excellent memoir has described the conversion of pus-cells into a fibrinous clot, thus confirming an observation of Rovida.

Were it possible to bring about this peculiar transformation, working with pure cells, and under the simplest possible conditions, we should doubtless gain, by an investigation of the process, a further insight into the structure and function of the colourless corpuscles. Some observations with this end in view have been made upon the leucocytes which are heaped up in the lymphatic glands.

My attention was directed to these, since they are always to be easily obtained in considerable quantities and in an almost pure condition. I believe my observations are the more valuable as I have considered the influence of the living blood, of the plasma and of the serum on these cells.

This part of the research was rendered possible by using the blood of dogs which had, shortly before this was drawn, received an intravenous injection of peptone.

Although the time at my disposal has only allowed me to answer a small part of the questions which present themselves in this subject, and can be solved by the means I have used, yet I am in hopes that the results I have obtained will afford a sure starting-point for future inquiries.

1. In the preparation of pure leucocytes I have made use of the lymphatic glands of recently killed dogs and calves: the

glands were those of the neck, pelvis, and mesentery, and, in short, all the glands that seemed likely to afford a good supply of cells.

The glands were freed as much as possible of all adherent fat and connective tissue with the scalpel, carefully washed with 0·5 per cent. NaCl solution, cut up fine with scissors, the small pieces moistened with 0·5 per cent. salt solution and tied up in a small linen bag. The cells were then beaten out of this bag by means of a pestle into a mortar which contained a few c.c. of 0·5 per cent. salt solution. The pulpy mass thus obtained was mixed with about 100 c.c. of 0·5 per cent. salt solution, and the whole centrifugalised. After some hours the leucocytes had collected at the bottom of the vessel in the shape of a small cake, and the fluid above contained fat-granules.

Under the microscope it was evident that the precipitate consisted almost entirely of lymph-cells, the sole impurity being a few red blood-discs.

The purified cells were found to be available for coagulation experiments, not only immediately after their preparation, but also for twenty-four to thirty-six hours later, provided they were kept in the ice-chest covered with 0·3 per cent. salt solution. Beyond this limit, however, the leucocytes lose the property which now concerns us, viz., the power of forming a clot with water, salt solution and blood-plasma.

2. Clotting of the lymph-cells by salt solutions and distilled water.

If to the emulsion of cells in 0·5 per cent. salt solution we add so much of a concentrated solution of NaCl or MgSO$_4$ that the fluid now contains 3 or more per cent. of the salt, and then shake it, we see in the course of the next few minutes that the cells stick together to form a turbid clot, which is drawn out into sticky threads when raised out of the surrounding fluid by means of a glass rod. If this mass is transferred to distilled water, it becomes denser, and unfolds so as to form whitish pellicles about a square inch in extent.

Solutions of other neutral salts probably have a similar

action, but I have not yet made sufficient investigations on this point.

If the original 0·5 per cent. salt solution in which the cells are suspended be diluted with a large amount of water, clotting again takes place. In this case, however, the clots appear in the form of pellicles, so that the intermediate slimy stage which was produced by the salt solutions is absent.

It is instructive to add the salt solutions to the cells under the microscope. We then see how at first the cells shrink, sending out irregular processes, and become covered with granules which appear to flow away from them. When the movement has ceased, we see the nuclei of the cells intact; the cell-bodies, however, have disappeared, and instead of them we see a transparent mass, beset with granules, stretching from nucleus to nucleus and embedding these firmly, as can be easily recognised on moving the cover-glass slightly.

If the denser salt solution be now replaced by water, fine fibrils make their appearance between the nuclei.

By staining such a preparation, we can prove that only the body of the cell is transformed into the fibrinous structure, the nucleus remaining quite unaffected. In all cases, however, a few cells remain intact. To save recurrence to microscopical details, I may here mention that the fibres into which the cell-bodies are converted are much more evident when the disintegration of the lymph-cells is induced by means of peptone-plasma. In this case, indeed, the clot may be confounded with fibrin.

The clot which is produced by acting on the cells with a 3 per cent. salt solution never occupies the whole space which is filled by fluid.

If we allow the cells to remain standing in the solution for some time, and then pour off the supernatant fluid, it is impossible to detect any globulin in this, or to extract any from the jelly by washing with 1 per cent. salt solution.

3. Coagulation of the lymph-corpuscles by means of peptone-plasma.

If a living dog is injected with a solution of peptone (0·3 grm. to each kilo. of body weight) into the jugular vein, the blood drawn from it after the lapse of a few minutes remains fluid.

By means of the centrifugal machine a clear plasma may be obtained from this blood, which varies in its coagulability under different conditions. As a rule we can get a firm clot from a plasma which has been centrifugalised for some hours and then separated from the red precipitate, by passing a stream of CO_2 through it for some minutes, or by adding to it an equal volume of water.

As Fano has recently discussed these phenomena fully, I must refer to his paper on the subject.

If we centrifugalise this coagulable plasma again, until no further precipitate is produced, and then let it stand in a glass cylinder for twenty-four hours at 0° C., we get a plasma which in some cases cannot be made to clot either by CO_2 or by the addition of an equal bulk of water, or even by addition of para-globulin, fibrin ferment, or blood serum. At other times, how-ever, the CO_2 and the addition of water still act efficiently in inducing coagulation.

A plasma which remains entirely fluid even when kept for days, and will not clot with any of the aforementioned means, clots unfailingly in a few minutes through and through if purified lymph-cells be added to it, and diffused equally through all parts of the liquid by shaking or stirring. Fibrin which has been produced by this means is absolutely undistin-guishable in its physical properties from the fibrin obtained from normal blood; like this, it is dense and fibrous.

In chemical properties, too, it resembles ordinary fibrin save in one particular, viz., it does not swell up in 0·2 per cent. hydrochloric acid.

If we lift the cake, into which the plasma is converted, out of the vessel by means of a glass rod, a clear fluid drops from it and the fibrin shrinks to a small volume. This shrinking is much hastened if the escape of the fluid is aided by making a number of cuts into the mass with scissors.

If we repeat the experiment with clear fluid that has thus dropped out, adding to it another portion of purified lymph-cells, a second coagulation takes place. The same experiment can be performed for a third time with the fluid that has run out of the second fibrin cake. At the fourth or fifth time, however, the cells no longer act, and henceforward both they and the fluid to which they are added remain unaltered.

When the property of the plasma to destroy the cell-bodies is thus exhausted, the fluid is equivalent to ordinary serum, which also has no power to alter lymph-corpuscles.

Since in the variety of peptone-plasma just described, the clotting could only be induced by the addition of lymph-cells, it was possible to decide whether only the mass of cells added, or some constituents of the plasma in addition, were represented in the fibrin formed. This may be arrived at in two ways. I shook a considerable amount of purified lymph-cells in 0·5 per cent. salt solution till they appeared to be evenly distributed, and then divided the whole fluid into two equal parts. One part served to estimate the weight of the dried lymph-cells, the other I added to 20 c.c. of peptone-plasma of the variety I have just described, and after five minutes, when the clotting was complete, I took out the fibrin and washed, dried, and weighed it. The weight of the added cells was 0·17 grm., that of the fibrin 0·21 grm. If we take into account the experimental error which is unavoidable in such a mode of determination, especially as a perfectly equal distribution of the cells and a thorough purification of the fibrin must be difficult to attain, I think we are justified in concluding that no weighable constituent of the plasma has taken part in the formation of the fibrin.

The following may serve to illustrate the method used in determining the weight of the cells and fibrin :—

Cells.—The salt solution, containing the cells in suspension, was first centrifugalised. The clear supernatant fluid was siphoned off from the precipitated cells, which were then washed out of the glass on to a weighed filter by means of alcohol.

The cells, which all remained on the filter after this treatment, were then washed with dilute alcohol to remove the sodium chloride, dried, and weighed. The fluid that had run through the filter was evaporated to dryness on a water-bath, then extracted with absolute alcohol, and the resulting solution filtered and evaporated to dryness.   The weight of the residue thus obtained was added to that of the washed cells.

Fibrin.—The clot was lifted up with a glass hook, in order that the serum might drain away as completely as possible.   It was then covered with 0·5 per cent. NaCl solution, pressed out repeatedly, and finally washed with salt solution till the washings showed no trace of proteid.   The salt was then removed by washing with distilled water, and the residue dried at 100° C., and weighed.

The quantitative estimation just described has left some doubt whether a part of the substance forming the clot might not be derived from the plasma.   To decide this question, the amount of coagulable proteid in the plasma was determined before and after clotting had been induced by the addition of lymph-cells.   Since the cells were added to the plasma suspended in 0·5 per cent. salt solution, the dilution of the fluid thus produced must be taken into account, and so the volume of the salt solution added must be known.   The proteid was precipitated by neu-tralising and heating, and then washed with water and hot alcohol.   100 parts of the plasma contained 6·76 parts of proteid; the same quantity of peptone-serum—*i.e.* the fluid which had drained away from the fibrinous cake produced by the addition of lymph-cells—contained 6·30 parts of proteid.

As the plasma suffers no loss in proteids when the cells in it are transformed into fibrin, the only part it can play in the process of clotting must be similar to that played by salt solu-tions, which bring about the transformation of the cells into fibrin.   This property, however, is not possessed by plasma to an unlimited extent.   After the plasma has brought about the de-composition of a certain quantity of cells, it has no effect on any further portions of cells which may be added to it.

The most obvious explanation of this would be that in the production of a clot the substance which was able to bring about the conversion of the cells is either removed or decom-posed.   If this substance has become part of the clot, its quantity must be regarded as so small that it makes no appreciable difference in the weight.

From this peptone-plasma, which is only coagulable by the addition of cells, we must distinguish another variety, which can be made to afford a clot by leading through it a stream of CO_2 or by the addition of an equal volume of water. This sort of plasma also has the power of converting the cell-bodies of the lymph-corpuscles into fibrin. When this conversion has once taken place, the plasma not only loses its effect on lymph-cells, but also cannot be made to give any further clot by means of CO_2 or the addition of water.

Hence it seems extremely probable that those substances which would be precipitated from the plasma by means of CO_2 or addition of water, take part in the formation of the clot. I cannot, however, bring forward any facts which would decide this question.

4. Behaviour of the living circulating blood with lymph-corpuscles.

After it had been shown that the plasma of peptone-blood, which had been obtained by bleeding, was quite different from the serum of this and of ordinary blood in its behaviour with lymph-cells, it became important to know in which category we should class the intravascular circulating plasma. According as this did or did not cause a destruction of the lymph-corpuscles which were added to it, our judgment of the properties of extra-vascular peptone-plasma would vary.

To decide this, three forms of experiment were chosen. The first mode of procedure consisted in ligaturing the jugular vein of a large dog just before its entrance into the thorax, dissecting out its two divisions below the submaxillary gland, putting a temporary ligature on one of them, and then injecting, with the usual precautions, a number of lymph-corpuscles suspended in 0·5 per cent. salt solution into the other.

The jugular vein, which had already been emptied of blood, appeared after the injection as a bright red turgid cord. After the aperture through which the injection had taken place had been closed, the temporary ligature on the other branch was loosened ; the blood from the head immediately rushed into the enclosed part among the mass of cells, which could not escape on account of the ligature of the main trunk. Half an hour later the animal was killed by curare, and the vein which served for the experiment was carefully dissected out.

During this operation it could be already observed that the contents of the vein were fluid, and this was fully confirmed as they were emptied out into a glass vessel. After the blood had been emptied out it clotted as usual.

If the conversion of the lymph-cells took place in the living circulation exactly as it does in extravascular peptone-plasma, the introduction of these in large quantities into the heart would certainly bring about a great disturbance in the health, if not the death of the animal.

I have performed experiments of this nature on three dogs : in two of them peptone was injected into the jugular vein before the introduction of the cells. Although in the animal which had received no injection of peptone 25 c.c., and in the other two cases 30 c.c., of a stiff mass of cells were introduced into the heart through the veins of the neck, yet none of them suffered any appreciable bad effect from the injection of the cells, and when they were killed afterwards, by opening the carotid, the blood was perfectly fluid.

The dog which had received no peptone I kept alive for four hours after the injection of the cells ; its blood clotted after leaving the vessels as quickly as usual.

The dogs which had received first peptone and then lymph-cells I dared not keep alive so long, if the peculiar properties of peptonised blood were to be recognised. I therefore killed one a quarter of an hour, the other half an hour after the cell injection, by opening the carotid. The blood thus obtained remained completely fluid for hours. At the autopsy, which was carefully performed on the animals, not a trace of embolism was to be found, even in the lungs. Only in one case did the lungs present an occasional red spot, which, however, disappeared immediately on pressing them with the finger.

No more convincing proofs than these observations could be

given, that within the living blood-stream the conversion of lymph-cells into a substance similar to fibrin does not go on, whether the blood is normal or has received an injection of peptone.

Against my experiments it might be objected, though with little *a priori* probability, that the shed blood of the animal which had been peptonised was an exception to the general rule, and had not the power of changing the cell-bodies into a clot. To meet this objection, a small portion of the cells which were to be injected into the heart was placed in a watch-glass. Some minutes after the injection of peptone was completed, I allowed a small quantity of the blood to flow into the watch-glass. After a short time the whole had clotted firmly. Thus the same blood which within the vessels had no effect on the lymph-cells, outside the vessels destroyed them to form a clot.

5. I may sum up the results of my experiments on the manner and causation of the destruction of the lymph-cells as follows :—

The proteids which are present in the cell-bodies of the lymph-corpuscles contain all that is necessary for the formation of a fibrin which resembles blood-fibrin in many particulars.

In the conversion of the cells into fibrin only chemical conditions come into play, since it takes place in cells whose protoplasm has long lost all power of movement, and is therefore dead.

Among the substances which can bring about this conversion of the cells is one of which we know little, save that it is a constituent of blood which has left the vessels but has not yet clotted ; for this cell-destroying substance is present neither in living intravascular blood nor in serum of clotted blood. It cannot, therefore, be identical with the fibrin ferment discovered by Alexander Schmidt. The dead cells which have been injected into the living blood evidently alter their characters ; for the blood of the animals which had been injected with cells and peptone must have clotted after it left the vessels, had not the cells themselves been altered.

As, however, the presence of the lymph-cells in the shed

H

blood is only deduced from the fact that they cannot, on account of their coalescence, wander out of the vessels, and as there is no sign that they have settled down in any of the capillaries, it is still desirable to prove directly the presence of the injected lymph-cells in the shed blood.

It will be possible to say more about the relation of fibrin of blood to that produced from the lymph-cells when the researches on normal clotting have led to a thoroughly satisfactory conclusion. At present we must regard the fibrinous clot resulting from lymph-cells as different from blood-fibrin, since the latter swells up in 0·2 per cent. HCl, while the former shrinks in this, similarly to mucin.

We find still another difference in the weight. The weight of the fibrin which is formed from blood does not consist merely of the weight of the disintegrated leucocytes, it also includes a derivative of a globulin that was present in the liquid plasma, viz. fibrinogen.

In direct distinction to this, the clot formed from lymph-cells consists only and entirely of the substances present in these. Since the weight of the clot was found equal to the weight of the lymph-cells from which it was produced, we have no grounds for assuming that one constituent of the plasma has entered into the formation of the clot, and another constituent of the cells has been given in exchange to the liquid.

There is thus a difference not merely between the fibrins formed from the two kinds of cells, but also between the cells themselves; this is shown by their different manner of disintegration and the different degree of tenacity with which they maintain their shape.

Solutions of $MgSO_4$ destroy the framework of the leucocytes of the blood and of the lymph-cells, and in each case a fibrinous clot is obtained. In the clot produced from the cells, however, the only remnant of their former structure is seen in the presence of nuclei; whereas after the breaking up of the leucocytes of the blood, between the fibrinous masses there still lie other masses resembling protoplasm.

Still more striking are the different powers of resistance shown by the two kinds of cells to peptone-plasma.

If ordinary coagulable blood be allowed to flow out of the carotid artery into peptone-blood or peptone-plasma, the mixture remains fluid (as Fano showed), provided that it contains about equal volumes of each ; the colourless cells of the coagulable blood are thus not destroyed. The lymph-cells, however, quickly undergo disintegration when added to the shed peptone-blood.

FURTHER OBSERVATIONS ON THE COAGULA-
TION OF THE BLOOD [1]

In a previous communication on the relation of the white blood-corpuscles to the coagulation of the blood,[2] I pointed out that there are essentially two processes to be considered.

The blood-plasma, after it has left the vessels, exerts an active destructive influence on the white cells, whereby the latter are themselves converted into fibrin, but at the same time a certain substance (or substances), which I then, in accordance with the usual doctrine, called 'fibrin ferment,' is liberated from the cells.

This substance (or these substances) is able to bring about the coagulation of the fibrinogen in the plasma.

Previous to my communication, the active destructive power of the plasma had been entirely overlooked. Writers on the subject always speak of a breaking up (*Zerfall*) or death (*Absterben*); there is never the slightest hint that the plasma plays an active part in the matter.

Rauschenbach,[3] under the direction of Prof. Alexander Schmidt, has repeated and extended my experiments, using cooled plasma instead of peptone-plasma. He comes to exactly the same conclusion as I do with regard to the active destructive power of the plasma, except that he does not make the distinction which I do between plasma in the vessels and plasma which has left the vessels. He, with, I think, perfect right, makes much of this action of the plasma (*spaltende Wirkung*), but he

---

[1] [From the *Journal of Physiology*, vol. iv. No. 2, Aug. 1883, p. 226.]

[2] *Proc. Roy. Soc.* vol. xxxii., 1881, p. 413. [Contents identical with those of the third part of the preceding paper.—Ed.]

[3] 'Blutplasma und Protoplasma,' Inaugural Dissertation, Dorpat.

does not seem to be fully convinced that I had discovered the fact long previous to his communication.

In extending my own researches, I first directed my attention to the body or bodies which are separated from the white cells, and which induce coagulation of the fibrinogen of the plasma ; and in doing so I have come upon a fact which is of the very highest importance, not only for the question of the coagulation of the blood, but for the very much greater question of the nature of the chemical processes in protoplasm which constitute life.

This fact is that lecithin, a body omnipresent in protoplasm, can bring about coagulation.

I describe now the experiments on which this statement is based.

The experiments were performed on dogs' blood which had been prevented from coagulating by injection of peptone.

For the satisfactory carrying out of the experiments in question the peptonisation must be very complete, the plasma from the blood must be centrifuged until absolutely no further sediment is obtained. Such a plasma is not coagulated by passing through it a stream of carbonic acid, no matter how long or how frequently this may be repeated, nor is it coagulated by adding other acids, e.g. acetic, till a slightly acid reaction is present.

But although a stream of carbonic acid or the addition of another acid does not induce coagulation, it does bring about a certain change in the plasma, as will be apparent from the following experiment.

Exp.—Peptone-plasma uncoagulable with CO_2. To one part plasma one part normal serum is added. After twenty-four hours a scarcely perceptible clot ; no further increase on standing twenty-four hours longer.

To one part of same plasma, but one through which a stream of CO_2 had been passed, one part normal serum is added. In ten minutes complete coagulation has occurred, so that the vessel can be inverted without anything falling out.

We see, therefore, that a plasma practically incoagulable with serum is rendered easily coagulable after a stream of carbonic acid is passed through it. The plasma in question is totally uncoagulable with fibrin ferment, but it becomes readily coagulable with fibrin ferment after a stream of carbonic acid has been passed through it. It is not necessary to use carbonic acid. Neutralisation with acetic acid acts just in the same way.

Plainly, therefore, in peptone-plasma either fibrinogen is not present as such (it is present, for it can be obtained by the salt method), or there is something present which prevents its coagulation. The acidification does away with these obstacles, and when the necessary additions have been made coagulation occurs. To this point I will return later.

I will call to mind that I am speaking of a plasma which is not coagulable with acids.

If to such a plasma lymph-cells obtained in the method previously described by me be added, coagulation occurs. If sufficient lymph-cells be added, the coagulable substance in the plasma disappears, that is to say, you get a serum. This serum will bring about coagulation in a further portion of plasma.

Now the usual doctrine about cells is that they give out ferment and paraglobulin.

But it is quite evident they must do more than this, for, as we have seen above, normal serum which contains both produces by itself alone a very faint coagulation. But it always does induce a certain amount of coagulation, which is as it ought to be, since normal serum, of course, contains the products of the disintegration of white cells, although to a much less extent than the plasma I have been talking about.

Both the cells and the serum from the coagulation brought about by cells act with very great rapidity and completeness. They must, therefore, give out something which exerts a similar influence to that exerted by the passage of a stream of carbonic acid. What this influence is I do not at present know, but I have learnt something as to what is the body or what are the bodies which the cells give out.

For I find that the *alcoholic extract* of the cells acts just as well as the cells themselves. Now the alcoholic extract has invariably an acid reaction, and, as will be remembered, we are now dealing with a plasma not coagulated with acids but rendered coagulable by acids. The acid of the extract plays a part, but it is not sufficient for coagulation; the other substance in the alcohol extract is necessary, and this other substance is lecithin.

The alcoholic extract is prepared as follows:—

The lymph-cells are extracted with hot alcohol; this is filtered off and allowed to cool. A precipitate occurs on cooling, and this is again filtered; the filtrate is evaporated to dryness. The residue is treated with a little cold absolute alcohol; a very large portion is left undissolved. After filtration the clear alcoholic solution is evaporated to dryness. It is then dissolved in cold absolute ether, and this is filtered and evaporated. The ethereal solution has an acid reaction. One portion of this extract is used for coagulation experiments, the other is used for analytical purposes.

The coagulation experiments are made in the following manner:—

A portion of the extract is rubbed up to a paste or thick emulsion with a drop or two of dilute sodium carbonate. This neutralises the acid present. On diffusing this emulsion through a portion of plasma no coagulation results. But when I pass through this plasma a current of CO_2 complete coagulation occurs in from ten to twenty minutes.

The plasma without this emulsion is totally uncoagulable with CO_2.

Now as to the chemical nature of this extract which can bring about coagulation: the residue from the ether solution is not crystalline, but forms a yellowish waxy mass. In water it is not soluble, but it swells up, and if examined under the microscope the formation of the peculiar myelin drops, characteristic of lecithin, is observed with great distinctness. If a portion be incinerated with sodic carbonate and saltpetre, it leaves an ash very rich in phosphoric acid. If it be dissolved in a little alcohol, and to this be added an alcoholic solution of platinum chloride, a voluminous yellowish-white precipitate is caused. This precipitate is not distinctly crystalline, and is very easily soluble in chloroform. It contains platinum, chlorine, and phosphorus. The filtrate from the platinum chloride precipitate, when freed from superfluous platinum by a stream of H_2S, and evaporated to dryness, leaves a comparatively very small residue, which melts on the water-bath. This is easily soluble in ether and alcohol, the solutions having an acid reaction. It is not soluble in water, but is soluble in dilute alkalies. If the alkaline solution be acidified and heated, oily drops appear on the surface. If the substance be treated with concentrated caustic soda, a jelly-like mass is the result. In fact, the residue in question consists of fatty acids.

Now if this residue be treated with a little carbonate of soda, and the influence of the resulting soap on coagulation be tested in the manner above described, it is found that it is absolutely without any influence.

The alcohol-ether extract which brings about coagulation in the manner above described consists then chiefly of lecithin; besides the lecithin there is a small quantity of fatty acids. When the lecithin has been removed, the extract has lost its power of bringing about coagulation.

A peptone-plasma which does not coagulate with CO_2 is a plasma in which there are no white cells and no products of the breaking up of white cells; and we have seen that the addition of lecithin, which is abundantly present in the cells, renders the plasma coagulable under the above-mentioned circumstances.

I will only remark, although it is scarcely necessary for me to do so, that the alcohol-ether extract certainly does not contain any fibrin ferment. Heating it to 100° C. in water does not destroy in any way its activity.

The ordinary Schmidt's ferment is free from lecithin.

I will call to mind certain well-known facts in coagulation. The transfusion of serum or defibrinated blood is, as a general rule, not followed by thrombosis. Yet a solution of fibrinogen coagulates at the temperature of the body with great rapidity. This has always been a point of great difficulty in connection with the coagulation doctrine. The peptone-plasma behaves, in this respect, just like the plasma in the vessels. The peptone-plasma behaves towards ferment just as if it contained no fibrin-ogen. After a current of CO_2 has been passed through it, it behaves just like a solution of fibrinogen.

It is on evidence of this nature that we talk of zymogen in gland-cells. And I think it is almost admissible to talk about a mother substance of fibrinogen in the blood.

The able researches of Hammarsten have shown that one albuminous body is sufficient for fibrin formation.

But Hammarsten also finds that, when a fibrinogen has been dissolved and reprecipitated a great number of times, *special* ferments are necessary for fibrin formation; and it is a very suggestive circumstance that these special ferments lose their activity after standing some little time under alcohol. (Removal of lecithin.)

I need hardly add that the above brief and condensed state-ment is intended only as a preliminary communication.

ON THE COAGULATION OF THE BLOOD [1]

It is now well known that the coagulation of the blood is brought about by an interaction of blood-plasma and leucocytes. It can be prevented by injecting peptone into the blood of a dog, apparently only for the reason that the action between cells and plasma cannot take place.

The grounds on which this assertion is based are as follows. Fano [2] found that fibrinous flakes occurred in the peptone-plasma ; a considerable breaking up of leucocytes was likewise observable. In my work on the chemistry of the blood-corpuscles [3] I have shown that the weight of the white cells in the peptone-blood is not only greater than that of the cells in the corresponding amount of whipped blood, but also greater than that of the cells in the same amount of magnesium sulphate blood. I have also shown that, if leucocytes from lymph-glands be added to the peptone-plasma, coagulation invariably occurs, which can only be ascribed to the action of the cells.

The different specimens of plasma that one may get from peptone-blood are not quite identical in their behaviour, and may be divided into two sorts. The difference between them will be apparent as we proceed.

The blood, immediately on being drawn off, is put on the centrifuge for about six hours. This period is long enough to remove all the red, but not nearly all the white corpuscles. The clear plasma is separated from the corpuscles and kept in ice until the next day, when it will be found to have lost its clearness. It is now either cloudy or with a more or less floc-

---

[1] [Translated from *Du Bois' Archiv*, 1883, p. 389.]

[2] ' Das Verhalten von Pepton,' *Du Bois' Archiv*, 1881.

[3] ' Zur Chemie der Blutkörperchen,' *Du Bois' Archiv*, 1881. [Collected Papers, p. 87.]

culent coagulum in it.   In either case it is coagulable if treated
with CO_2, or if diluted with water.

Let us first consider the plasma which was only cloudy.   It
is again centrifugalised until no further sediment at all is
formed.   (This sediment consists of white cells and faintly-
coloured *débris*.)   The plasma will be now found to have
entirely lost its power of coagulating with CO_2 and on dilution.

We will now confine our attention to this plasma.

Behaviour of the plasma towards fibrin ferment.—The plasma
will not clot with fibrin ferment prepared according to Alexander Schmidt's method, and which is very effectual on salt-
plasma.   But if it be first submitted to the action of a stream
of CO_2 it coagulates very readily with ferment.   The plasma
is not absolutely uncoagulable with normal dogs' serum, although
never more than a few flakes of fibrin are formed, and these
may appear rather quickly, *i.e.* within an hour, or after a much
longer time.   If, however, the same amount of serum be added
to a plasma after it has been treated with CO_2, complete coagulation will occur very rapidly—within ten minutes.

Although CO_2 does not of itself initiate coagulation, it
causes certain changes in the plasma which render the latter
coagulable with ferment and with serum.   The same effect is
produced with other acids, as well as with carbonic acid, such,
for instance, as acetic acid.

Action of leucocytes.—If leucocytes from lymph-glands be
added, clotting occurs rapidly after about five minutes.   Provided that sufficient lymph-cells have been employed, all coagulable matter disappears from the plasma.   The serum from this
clot, when freed from cells, induces coagulation in a fresh portion
of plasma.   It follows that the cells must yield one or more
substances which are capable of causing the coagulable matter
in the plasma to clot.   It is well known that Alexander
Schmidt asserts that the white corpuscles give fibrin ferment
and paraglobulin.   But they must do more than this, since we
have seen that serum which contains both is very little effectual.
They must also yield a substance which acts in a similar manner
as the passage of CO_2 or the addition of an acid.   So far, I

have not obtained any knowledge concerning the nature of this substance.

Whether the white cells really give paraglobulin and ferment I will not venture to decide. But, however this may be, a body may be produced from the white cells which is free from paraglobulin and ferment, and has nevertheless the power of setting up coagulation. The addition of an alcohol and ether extract of the cells acts as well as the addition of the cells themselves.

The alcoholic extract of cells is thus prepared. The cells are boiled with alcohol; the latter is filtered off and allowed to cool, when it is again filtered to separate the fats, &c., that have been deposited, and evaporated to dryness. The residue is extracted with cold absolute alcohol, a good deal remaining undissolved. The clear alcoholic extract is evaporated and extracted with cold absolute ether. After filtration, this solution, which always gives an acid reaction, is evaporated to dryness at a low temperature. The residue has the following characters. It forms a yellowish, waxlike, non-crystalline mass, which is soluble in cold alcohol and ether. It is insoluble in water, but swells up in it, and very beautiful myelin forms may be seen under the microscope. If incinerated with a little soda and saltpetre, it gives an ash containing a quantity of phosphorus. If dissolved in cold alcohol and a solution of platinum chloride in alcohol added, a flocculent yellowish-white precipitate separates out. This precipitate does not appear to be crystalline. It is readily soluble in chloroform, and contains platinum, chlorine, and phosphorus. If this precipitate be filtered, and the filtrate freed from platinum by means of H_2S and evaporated on the water-bath, we get a very small residue consisting of fatty acids. The ether extract consists, therefore, of lecithin, and of a small amount of fatty acids.

In order to try the effect of this extract in coagulation, we proceed as follows. A little of the dried substance is made into a thick paste with one or two drops of a dilute solution of carbonate of soda, which neutralises the acid. This paste is mixed with some peptone-plasma. Coagulation does not occur until a stream of carbonic acid is passed through, when it takes place completely in a few minutes. As a control experiment, some more of the same plasma is treated with CO_2, but, in spite of repeated treatment with CO_2, no clotting is found after twenty-four hours.

The extract of cells is therefore essential for the initiation of coagulation, and the effectual agent in the extract is the lecithin; for, as we have seen, lecithin and a small amount of fatty acids are the sole constituents of the extract, and coagulation invariably occurs if the experiment be carried out in the way just described.

After the lecithin has been removed from the substance thus prepared, the remainder has absolutely no effect in coagulation. Lecithin is, therefore, a factor in this process. If the extract is added to the plasma, without previous neutralisation of the acids, coagulation begins without extraneous aid—i.e. without the help of carbonic acid. The acids present in the extract act, therefore, in the same way as the carbonic acid.

In this experiment the lecithin must be intimately mixed with the plasma to ensure a successful result.

I may add that the extract from cells does not lose its efficacy when boiled with water. I mention this to avoid the objection that ferment might possibly be present, although this is rendered highly improbable by the method of preparation. I have substituted similar experiments with lecithin from other sources, especially from red blood-corpuscles. This latter variety acts in the same way as the lecithin of the white cells. I have also tried lecithin prepared from eggs, according to Strecker's method. But although successful in a few cases, yet in most cases this variety had no power to initiate coagulation. The ineffectual preparations were invariably observed to contain neurine in considerable quantity, and their inefficacy may have been due to the presence of this body. To obtain a thoroughly reliable extract, the lecithin must be prepared from leucocytes or from red blood-corpuscles.

We can get a precipitate of proteid in peptone-plasma by the addition of powdered salt. When the precipitate has been filtered, pressed out and dissolved in water, the solution will not clot of itself, but does so on addition of Schmidt's ferment. This precipitate is, however, very rich in lecithin.

It is absolutely certain that the addition of paraglobulin to any liquid containing fibrinogen may greatly increase the amount of fibrin formed. But as the paraglobulin precipitates obtained from serum are always impure from the admixture of lecithin, it is at least open to question whether the paraglobulin or the lecithin promotes the formation of fibrin. Rauschenbach's [1] observations, that yeast and spermatozoa, which are both rich in lecithin, can produce coagulation, are in harmony with my results.

We have seen that peptone-plasma does not coagulate spontaneously with ferment, and but very slightly with normal serum. In this respect it resembles the plasma within the vessels. For the transfusion of whipped blood from the same animal seldom leads to thrombi of any extent, and the injection of a solution of fibrin ferment, which acts powerfully on salt-plasma, will not, according to Alexander Schmidt, cause intra-vascular clotting.

Hitherto we have been dealing with a peptone-plasma which is not coagulable with CO_2. This property is only possessed by the plasma from which all leucocytes have been removed by the centrifuge before the slightest coagulum makes its appearance. If this has happened to any extent, the plasma retains its power of clotting with CO_2 even after repeated centrifugalising. In fact, a perfectly pure plasma can no longer be obtained; the products of the breaking down of cells are mixed up with it, for there is no coagulation without the breaking down of cells.

My experiments tend to show that plasma contains not fibrinogen as such, but a substance from which it may arise.

[1] 'Blutplasma und Protoplasma.' Dorpat.

CLASS II

THE DISCOVERY AND DEMONSTRATION OF
THE OCCURRENCE OF COAGULATION
WITHOUT FERMENT ACTION

ON THE COAGULATION OF THE BLOOD[1]

THE following communication is a continuation of the results already published by myself as to the influence of lecithin in producing coagulation of the blood.[2]

The plasma used in those experiments was peptone-plasma. In the present case I have made use of blood which has been prevented from coagulating by being, immediately after leaving the body, cooled down to a temperature of about 0°.

For experiments on cooled plasma it is best to use the blood of the horse. I was, however, unable to obtain any horse's blood, and therefore employed dog's blood. In horse's blood the corpuscles sink rapidly and the coagulation is very tardy; it is hence easy to obtain plasma. This is not the case with the dog, and I was therefore led to adopt a particular method of experimenting.

The coagulation of the blood is brought about by a certain interaction of the white corpuscles and the plasma. By rapid cooling this interaction is to a great extent suppressed, and hence the blood does not coagulate. But if one of the substances of which the white cells are made up be diffused through the cooled blood, coagulation does occur. This substance is lecithin.

The method of experimenting is as follows :—The blood is taken from a large artery and flows into a thin cylindrical metal vessel of about ½-inch diameter. This tube stands in a large vessel filled with broken ice and a small quantity of water to fill up the interspaces.

---

[1] [From the *Journal of Physiology*, vol. iv. No. 6, Feb. 1884, p. 367.]

[2] 'Further Observations on the Coagulation of the Blood.' *Journal of Physiology*, vol. iv. No. 2, 1883. [Collected Papers, p. 100.] 'Zur Gerinnung des Blutes,' *Du Bois' Archiv*, 1883. [Collected Papers, p. 106.]

I

For each experiment two such tubes are used.   Each holds
40 c.c.   In the one is contained 15 c.c. of ·6 per cent. NaCl
solution, in the other a similar quantity of normal salt solution
through which finely emulsified lecithin is diffused.

The blood flows directly from the artery into the metal
vessels, one vessel being filled immediately after the other.
Absolutely no interruption to the flow must occur.   The first
small quantity of blood is not used.

The following examples will show clearly the nature of the
results:—

I. *Large Dog.   Blood from Femoral.*

Tube 1 contains lecithin.

Tube 2 contains simple salt solution.

By means of glass rod mixing is carried out in each tube, a
separate rod being used for each tube.

Tubes filled at 4.15 P.M.

At 4.45 P.M., as is ascertained by feeling with glass rod in tube
1 (lecithin), the blood is firmly coagulated.

At this time the blood in tube 2 is perfectly fluid.   The temperature in the two tubes is the same, 2° at upper, 3° at lower part
of tube.

On removing tube 1 from ice and inverting, a solid clot, forming
a complete cast of the tube, slips out.

At 5.30 P.M. blood in tube 2, on trying with glass rod, is
evidently completely fluid ; on taking it out of ice and inverting,
the blood flows out ; it is perfectly free from coagula ; there are no
traces of coagula on the walls of the tube.

II. *Similar arrangements as in* I.   *Large Dog.   Blood from Carotid.*

Tube 1, lecithin.

Tube 2, simple salt solution.

Tubes filled at 4.3 P.M.

At 4.10 P.M. tube 1 completely coagulated.

　　　　„　　tube 2 completely fluid.

Temperature in the two tubes equal, 2·5° in middle of tube.

At 4.45 P.M. tube 2 still quite fluid.

At 5 P.M. imperfectly coagulated.

III. *Arrangement the same. The lecithin is badly emulsified, and
in flocculent pieces, which tend to rise to the surface of the NaCl
solution.*

Time of filling the two tubes 4.50 P.M.

At 5.30 P.M., tube 1 (lecithin), several loose coagula can be
fished out. Tube 2 apparently quite free. Temp. tube 1, 2·5°
bottom, 1·5° top ; temp. 2, 4° bottom, 2½° top.

On emptying tube 1 it is found to contain loose clots and fluid
blood ; it is coagulated firmly at the top so that it can be inverted.
The walls of the tube covered with clot.

At 6 P.M. tube 2 emptied. The contained blood is perfectly
fluid. No trace of coagulation on the walls of the tube.

I have made many such experiments, and always with the
like result; that is to say, the addition of lecithin causes
coagulation.

The lecithin used is prepared from lymph-glands in the
manner I have described in the above-quoted papers. It has a
slightly acid reaction and is not perfectly pure. It is rubbed
up with a drop or two of dilute Na_2CO_3 solution to a paste, and
this is diffused through the salt solution. The salt solution,
after the addition of the lecithin, is either neutral or very faintly
alkaline. It should be an emulsion as opaque as milk if the
coagulation is to occur quickly.

In my experiments on peptone-plasma I have shown that
the small quantities of impurities contained in the lecithin
preparations are of no effect in producing coagulation, and also
that lecithin from other sources is active, and hence I have felt
justified in speaking of the alcohol-ether extract of the glands as
lecithin, though it is not perfectly pure. From its mode of
preparation it is evident that this extract cannot contain any
paraglobulin.

In discussing the results I have obtained on peptone-plasma
with friends whose opinions are of worth, there has always been
a tendency to suppose that the results obtained are due to
presence of fibrin ferment in the lecithin preparations. In fact
this is, so far as I can see, the only objection one can raise. It
is, however, certainly not a valid objection. The lecithin is not

influenced in its activity by being boiled with water. I have tried this several times both with peptone and cooled blood, and, as a matter of fact, it has no fermentative activity.

There are many coagulable fluids which do not coagulate in the slightest on the addition of lecithin, but do so readily with fibrin ferment, although the lecithin in question acts perfectly with plasma. Such fluids are human pericardial fluid, hydrocele fluid so far as I have examined, and solution of fibrinogen prepared according to the salt method from peptone-plasma.

The experiments described were carried out in the Pathological Institute in Berlin. I am much indebted to Prof. Salkowski for the use of his laboratory.

Dec. 6, 1883.

CLASS III

THE TRUE NATURE OF FIBRINOGENS

ON THE ORIGIN OF THE FIBRIN FERMENT[1]

THE 'fibrin ferment' which makes its appearance in shed blood is generally, I believe, supposed to arise from the cellular elements of blood, either from ordinary white corpuscles or from some special kind of corpuscles, the cells so concerned discharging the ferment into the blood or setting it free by their actual disintegration. Without wishing to deny that this may be one source of fibrin ferment, I am able, I think, to bring forward evidence that ferment may make its appearance in blood-plasma perfectly free from cellular, and indeed from all formed elements, in which case it must arise from some constituents of the plasma itself, and not from cells of any kind.

It will be most convenient, perhaps, if I state the facts which I have to bring forward in connection with two series of experiments.

I. A measured quantity of blood was received directly from the carotid of a dog into a vessel containing an equal bulk of a 10 per cent. solution of common salt, great care being taken that the complete admixture of the blood and salt solution was effected as rapidly as possible. By the help of the centrifugal machine plasma was separated from this 'salted blood,' and this plasma was again subjected to the action of the machine until all traces of formed elements were removed. As is well known, a portion of such a plasma diluted with five times its bulk of water coagulates rapidly, whereas the undiluted plasma remains liquid for an almost indefinite time.

According to commonly received opinions, such a 'salted plasma' contains all the fibrin factors, including the ferment,

[1] [From the *Proc. Roy. Soc.* 1884, p. 417.]

the latter having already passed out of the cells into the plasma ; and the reason given for the absence of coagulation in such a salted plasma and its occurrence upon dilution is, that the presence of the salts presents a hindrance to the action of the fibrin ferment, and that this obstructive influence of the salt is removed by the dilution of the mass.

No one, however, as far as I know, has taken the trouble to ascertain whether fibrin ferment is present in such salted plasma. And, as a matter of fact, it is not ; whereas it does make its appearance as soon as dilution with water has taken place, as the following experiment shows :—

A portion of the undiluted salted plasma was treated with absolute alcohol in large excess, and the precipitate, after being allowed to remain under the alcohol for three or four weeks, was dried at a low temperature and extracted with water ; that is to say, the plasma was treated in the way usually adopted for obtaining a solution of ferment fairly free from proteids, &c. A portion of the diluted plasma, or rather of the serum resulting from the coagulation of the diluted plasma, was treated in an exactly similar manner.

The aqueous extract of the diluted plasma brought about coagulation in specimens of magnesium sulphate plasma (such as is usually employed for testing the presence of fibrin ferment) in from ten to fifteen minutes. The aqueous extract of the undiluted plasma brought about no coagulation in specimens of the same magnesium sulphate plasma even after the lapse of eighteen hours.

The conditions of each experiment were made as exactly alike as possible ; and the conclusion seemed inevitable that ferment is present in the diluted and coagulated plasma, but absent from the undiluted plasma.

This conclusion is, moreover, supported by the following experiments :—To a portion of the undiluted plasma above mentioned a small quantity of fibrin ferment was added, in the form of the dried precipitate thrown down by alcohol—*i.e.* a mixture of coagulated proteids and ferment. Coagulation took place. I have no record of the exact time elapsing between the addition of ferment and the appearance of the clot, but it was certainly not longer than three or four hours.

II. Of the so-called peptone-plasma (*i.e.* plasma of the blood of a dog after the injection of peptone into the veins, such blood, as is well known, coagulating with great difficulty), freed from all cellular elements by the centrifugal machine, two portions were taken.

To the one (A) a quantity of *lecithin* was added, the lecithin being rubbed up with the plasma so as to be diffused through it ; the other (B) was left untouched.

Through both a stream of carbonic acid was passed, with the result that while A clotted in about ten minutes, B after the lapse of half an hour showed no disposition whatever to coagulate. Both portions were then treated with excess of alcohol for the extraction of fibrin ferment in the usual way. The aqueous extract of A proved to be exceedingly rich in ferment, producing coagulation in magnesium sulphate plasma in about ten minutes. The similarly prepared aqueous solution of B produced no coagulation at all.

Now I have elsewhere,[1] in discussing the action of lecithin in promoting coagulation, shown that the coagulation which is brought about by the addition of lecithin is not due to the lecithin, or to any of its products of decomposition acting after the manner of a ferment, or to its carrying a fibrin ferment with it. In this case, therefore, as in the previous case of ' salted ' plasma, the ferment appears to be absent *before* coagulation, but to be present *after* coagulation.

I may here call attention to an observation made by Rauschenbach.[2] This observer found that the addition of yeast to plasma, prevented from coagulating by exposure to cold, brought about coagulation, and at the same time gave rise to the appearance of a large quantity of fibrin ferment. Nevertheless, he completely failed to extract any fibrin ferment from the yeast itself. Now yeast is very rich in lecithin, and it seems highly probable that the coagulation caused by yeast was due

[1] *Journ. of Physiol.* vol. iv. 1883, p. 226. [Collected Papers, p. 100.]
[2] ' Blutplasma und Protoplasma,' Inaug. Diss., Dorpat.

to the lecithin contained in it, and hence the appearance of the fibrin ferment after the addition of yeast, and consequent coagulation, is quite parallel to the result of the experiment with lecithin and peptone-plasma recorded above. In both cases the ferment appears to have arisen out of the plasma itself.

It is possible to obtain a coagulation in peptone-plasma without the addition of lecithin. For this purpose large dilution is necessary, followed by the passage of a stream of carbonic acid gas. But in such a case, however, coagulation is not only long in making its appearance, but the fibrin is formed, so to speak, in successive crops. Thus a feeble coagulation first appears, and if the clot so formed be removed, a succeeding coagulation is observed some time later, to be followed in turn by a third, and so on. When lecithin, on the other hand, is added, without previous dilution, the clotting is speedy and complete.

If the serum thus resulting from the coagulation of peptone-plasma brought about by large dilution and treatment with carbonic acid be examined for fibrin ferment in the usual way, it will be found to contain ferment, though much less than could be obtained from a corresponding quantity of the same plasma coagulated rapidly by the addition of lecithin. The relative amount of ferment appearing under different circumstances is illustrated by the following experiment :—

Of three equal portions of the same peptone-plasma, one portion was simply treated with a stream of carbonic acid gas, without any dilution, and did not coagulate ; a second was treated with a stream of the same gas after large dilution, and coagulated slowly ; to a third lecithin was added, and a stream of carbonic acid passed through it, with the result of producing a rapid and complete coagulation.

All three portions were treated in the same way for the extraction of the fibrin ferment, and the activity of the three aqueous extracts then prepared was tested under exactly the same conditions, with the help of magnesium sulphate plasma.

The first produced no coagulation after the lapse of twenty hours.

The second produced coagulation in four hours.

The third produced coagulation in five minutes.

The amount of ferment seems to be in proportion to the energy of coagulation ; and the presence of ferment after simple dilution and the action of carbonic acid gas shows that the ferment appearing after coagulation by the help of lecithin does not come from the lecithin itself.

Thus there is a remarkable coincidence between the occurrence of coagulation itself and the appearance of the fibrin ferment, and that in plasma freed most carefully from all cellular elements.

I believe, therefore, that I am justified in concluding that though fibrin ferment does not pre-exist in normal plasma, it may make its appearance in that plasma in the absence of all cellular elements, and must therefore come from some constituent or constituents of the plasma itself.

I am still engaged in investigations directed to find out what that constituent is, or what those constituents are.

ON A NEW CONSTITUENT OF THE BLOOD AND ITS PHYSIOLOGICAL IMPORT[1]

IN a paper on the Origin of the Fibrin Ferment, published in 'Proc. Roy. Soc.,' vol. 36, 1884, I showed that there exists, dissolved in the plasma, a body which can give rise to fibrin ferment.

I have proceeded with my investigations, and have succeeded in making some additions to our knowledge of this subject, which I here describe. As my researches are not complete, I confine myself to as brief an account as possible.

The subject is best studied in the blood of peptonised dogs; but, as I showed in the above quoted paper, similar results are obtained from normal salt-plasma, so that the results are not peculiar to peptone-blood. The body the presence of which gives rise to fibrin ferment can be isolated from peptone-plasma in the following very simple manner:—The plasma, having been completely freed from all corpuscular elements by means of the centrifuge, is cooled down to about 0°. The plasma, which was previously perfectly clear, becomes rapidly turbid, and after standing for some time in the cool, a very decided flocculent precipitate forms. I have already described this observation in a short note, ' Ueber einen neuen Stoff des Blut-Plasmas,' in Du Bois-Reymond's ' Archiv für Physiologie,' but it is necessary for me to allude to it here.

Now it is this body which gives rise to the fibrin ferment. So long as the former is present in considerable quantity, the latter clots readily on passing through it a stream of carbonic acid, or on dilution, and at the same time a very considerable quantity of fibrin ferment makes its appearance.

[1] [From the *Proc. Roy. Soc.* 1885, p. 70. (An account of this new constituent of the blood appeared also in *Du Bois' Archiv*, 1884.—Ed.).]

By prolonged cooling the greater part of this substance can be removed, and with its gradual removal the plasma clots less and less readily with CO_2, and less and less ferment is formed, till finally it becomes practically incoagulable—*i.e.* forms only a faint trace of fibrin after several days.   If some of the substance be again added to the plasma, it regains its power of clotting with CO_2.

(The substance must be added before it has stood very long: see under.)

It must be understood that the plasma, previous to the passage of the CO_2, is quite free from fibrin ferment, so that there can be no question of the ferment being mechanically removed by the precipitate.

Moreover, that it is really the body removable by cold which gives rise to the fibrin ferment, and not any second body which is mechanically carried down with the former, is shown by the fact that the diffusion of a large quantity of inert finely divided precipitate through the plasma, and its subsequent removal by the centrifuge, does not in any way do away with the power of the plasma to clot.

It is, therefore, justifiable to assume that when peptone-plasma clots readily and completely with CO_2, it must contain this new body in some quantity, and that when it will not clot, or only very imperfectly, after *repeated* treatment with CO_2 or dilution, this new body must be present in very small quantity.

Now I have found that the behaviour of peptone-plasma with CO_2 varies very considerably with the diet on which the animal is fed, and whether the animal is fasting or has been recently fed.   In some cases it clots readily, in others practically not at all.

Out of eight dogs fed on very lean meat only one gave a plasma which clotted at all fully, and in this case the clotting went on for two days.   From all the others the plasma, in spite of repeated treatment with CO_2, only gave rise, after two or three days, to a scarcely perceptible fibrin membrane.   The animals were killed about eighteen hours after the last meal.

Of six dogs fed on fat and meat for several days, all gave a plasma which clotted rapidly and fully in from twenty minutes to one hour after the CO_2 treatment. The animals were killed about eighteen hours after being fed.

Of two dogs fed on bread and meat, both gave a readily coagulable plasma.

One day's feeding on fat does not produce any effect; that is, the blood of a dog thus fed behaves like that of a dog fed on a lean meat.

A dog fed for some days on fat and meat was for five days previous to being killed put on fat alone; as a consequence it practically starved, as it ate scarcely anything. The blood from this dog clotted very incompletely.

Simple starvation for three days did away with the influence of fat in another case.

These results only hold good for dogs in health. In a dog with a suppurating wound, kindly placed at my disposal by Mr. Horsley, the plasma, in spite of a lean meat diet, clotted with very great rapidity, and contained an enormous quantity of the new body. All the other dogs were healthy, but were badly nourished when they came into my hands.

It is necessary for these experiments that the peptonisation should be complete.

For the better understanding of these results I must return to a further consideration of this new constituent of the plasma.

The turbidity which appears on first cooling the plasma, if examined microscopically, is found to consist of a great number of minute pale transparent bodies of a rounded shape, much resembling small organised bodies; such, for instance, as the stroma of the red corpuscles, except that they are of very various size, but generally much smaller than red corpuscles. They have a great tendency to run together into granular masses.

At first the precipitate is soluble on re-warming the plasma slightly, but it soon undergoes change, and loses the power of redissolving by heat. If the substance be collected by means of the centrifuge, it forms a disc or thin membrane at the bottom of the tube, much reminding one of fibrin, but closer examination shows that it presents marked differences from the latter, and that, in truth, it much more closely resembles the peculiar viscid body obtained by destroying leucocytes with dilute alkalies, &c.

On longer standing, however, it becomes in most cases still further changed, and is then undistinguishable from ordinary fibrin, swelling in dilute HCl like the latter. For further details as to the properties of this substance I refer to my paper quoted above.

We have already seen that this substance gives rise to fibrin ferment, but it does more than this in inducing coagulation.

Peptone-plasma is not coagulable with fibrin ferment. If we take some plasma rich in this new substance, and by means of CO_2 induce coagulation, we obtain, on removing the clot, a serum which has the power of inducing exceedingly rapid coagulation in a new portion of plasma, and this when the serum has regained its alkalinity. This serum contains ferment ; but, inasmuch as ferment is not sufficient to induce coagulation, it must also contain some other substance. Now leucocytes have exactly the same power. They give rise to ferment, but they also give rise to the other substance necessary for coagulation.

We see, therefore, that we have dissolved in the plasma a body exerting the same influence on the induction of coagulation as the leucocytes.

I think this is the strongest chemical proof that can be brought that the leucocytes break down to make, at any rate, a part of the proteid constituents of the plasma, and have shown above the influence which diet, &c., has on the extent of this process, a fact of obvious interest for the question of assimilation.

There is, however, another important conclusion to be drawn from these observations, viz., that one must admit, in addition to the ordinary fermentative fibrin formation, that fibrin may be deposited from blood by simple physical means, without any ferment process ; for this new substance becomes, as I have stated, true fibrin, and yet the plasma does not contain ferment. Possibly this mode of fibrin formation is of importance in the formation of a thrombus.

The peculiar microscopical characters should also be noted, as possibly affording an explanation of the observations made by Osler, Bizzozero, Hayem, and others. I refer, of course, to the granules, Blutplättchen, hæmatoblasts, described by these authors.

As I am actively engaged on this subject, and as I hope before long to produce a complete account of my researches on the coagulation of the blood, I have purposely confined myself to the briefest outlines.

ON THE FIBRIN-YIELDING CONSTITUENTS OF THE BLOOD-PLASMA [1]

THERE is no doubt that from every variety of blood-plasma a proteid body may be isolated, which can by appropriate means be converted into fibrin. This body, which is known as 'fibrinogen,' has been more especially studied by Hammarsten. This observer has shown that fibrinogen possesses characters which clearly distinguish it from the other supposed factor in coagulation, viz. paraglobulin, and also that solutions of fibrinogen will, when treated with fibrin ferment, give rise to fibrin. The only objection possible to Hammarsten's experiments is that the body which he isolated has either previously to or during the process of isolation undergone alteration; that it is, in fact, not the same body which is present in the circulating blood, but that it is, so to say, a sort of nascent fibrin. My observations bear on this point.

Peptone-plasma is obtained by injecting a solution of peptone into the veins of an animal and bleeding it directly afterwards. The blood does not clot, and by means of the centrifuge the plasma is obtained. The injection of peptone produces this effect by preventing the interaction of leucocytes and plasma which normally takes place in shed blood.[2] By repeated centrifugalising, the whole of the corpuscular elements can be removed from this plasma, and the pure plasma thus obtained can be made to clot in the most complete manner, giving rise to a large quantity of fibrin, and this without the addition of any further proteid body, so that the plasma must contain dissolved in it the mother substance or substances of fibrin.

[1] [From the *Proc. Roy. Soc.* 1885.]
[2] Wooldridge: ' Zur Gerinnung des Blutes,' *Du Bois' Archiv*, 1883, p. 380. [Collected Papers, p. 106.]

K

In a note presented to the Society a few weeks ago,[1] I described a new constituent of the plasma which gives rise to fibrin and to other bodies concerned in coagulation. I need not refer at length in the present paper to this new substance. It is separable from the plasma by cooling the latter, and after its removal the plasma still yields a large quantity of fibrin, and from this plasma, by Hammarsten's method, a body can be isolated agreeing in all particulars with Hammarsten's fibrinogen, and clotting readily with fibrin ferment.

The following observations refer to such a plasma in which the peptonisation is very complete, and from which the body separable by cooling has been removed.

Behaviour of the Plasma towards Fibrin Ferment and Serum [2]

In the vast majority of cases the plasma gives with either of the above only a very minimal clot, a few scarcely perceptible threads or membranes being the sole result of prolonged action. Serum is not more effectual than ferment.

But if, after the addition of serum or ferment, a stream or carbonic acid be passed through the plasma, or the plasma be diluted with several times its volume of water, it clots through and through, becoming quite solid.

The readiness with which coagulation takes place with CO_2 varies in different specimens, sometimes very rapidly, sometimes more slowly; sometimes it only occurs when the CO_2 treatment and dilution are combined.

There are two exceptions to the above statement:—

First. Sometimes neither ferment nor serum give by themselves the slightest trace of a clot in the plasma.

Secondly. They sometimes give a very considerable clot.

Both these exceptions are rare.

Now I take these experiments to show that the plasma con-

---

[1] 'On a New Constituent,' &c., *Proc. Roy. Soc.* 1884, p. 69. [Collected Papers, p. 124.]

[2] Serum from dog's blood. Ferment prepared from serum (dog's) according to Schmidt's method. The dry powder is added directly to the plasma, to avoid the effect of dilution.

tains a certain very small amount of true fibrinogen (coagulable with serum). In some cases even this small trace may be absent, in others it may be considerable. But the bulk of the coagulable matter of the plasma is not directly coagulable with serum. It is a body which is readily altered so as to clot with the serum or ferment. This alteration can be effected by dilution, by CO_2, or in the process of isolation. The body is not fibrinogen, but it readily passes into the latter.

I now turn to another kind of plasma which gives like results, but which is free from the objection that in peptone-plasma the proteids become altered by the peptone injection, an objection which I do not think to be at all justified by the facts. For the sake of convenience, I call it 'NaCl plasma.' It is obtained by receiving blood direct from the artery into a 10 per cent. solution of common salt, equal quantities of blood and salt solution being taken.

It is essential that the blood should be mixed with the salt solution with as little delay as possible. Very frequently the plasma obtained from this blood is a little stained with hæmoglobin.

Now if to this plasma ferment be added, a certain amount of clotting rapidly takes place; it is usually very inconsiderable. On very long standing no increase takes place. If, however, after the removal of the slight clot the plasma be diluted with four times its volume of water, it clots through and through.

Now 4 or 5 per cent. solution of salt does not interfere with the action of the fibrin ferment, and hence we must conclude that, as in peptone-plasma, so in salt-plasma, the bulk of the coagulable matter is not in the form of fibrinogen, but as a substance which must first be altered by dilution.

These conclusions are confirmed by the behaviour of the plasma on heating. As is well known, solutions of Hammarsten's fibrinogen coagulate on heating to 54–56° C.

If some NaCl plasma which has been treated with ferment, and from which the slight clot thereby caused has been removed, be heated to 56°, it remains perfectly clear; long exposure to this temperature does not alter it, and it can be heated up to a very high temperature, 90° and upwards, without the slightest coagulum forming, though at high temperatures it becomes opalescent.

The exact upper limit of coagulation varies; it is usually over 90° C. Of course, if hæmoglobin be present it interferes with the experiment. Now it will be remembered that this plasma, in spite of the removal of the small quantity of fibrin, contains a large quantity of fibrin-yielding matter.

The fact that ferment gives a slight clot in NaCl plasma may be taken as an indication that NaCl plasma does contain a certain small amount of true fibrinogen; and, in fact, if NaCl plasma which has not been treated with ferment be heated to 56°, it becomes turbid, and a slight coagulum forms. This, like the clot obtained by ferment, is sometimes very small indeed, sometimes more considerable.

Before describing the behaviour of peptone-plasma on heating I must make a slight digression.

In my note previously referred to [1] I showed that there exists dissolved in the plasma a body separable by cooling. So long as this body is present, coagulation is produced by passage of a stream of carbonic acid through the plasma, the whole of the coagulable substance being converted into fibrin. At the same time not only is fibrin ferment produced, but also a body capable of converting the coagulable body of the plasma into fibrinogen. To make this perfectly clear, I shall give an example.

Peptone-plasma, rich in the substance separable by cold, is treated with CO_2; it coagulates, the serum is pressed out from the clot and allowed to stand, when it again becomes alkaline. On adding some of this serum to a new portion of plasma, the latter clots completely and with great rapidity. The clotting is much more rapid than the original clotting with CO_2.

Now this serum contains fibrin ferment; but, inasmuch as ferment does not cause anything more than a trace of clot in peptone-plasma, the serum must evidently contain some special body which renders the plasma coagulable by ferment. It must contain a body capable of splitting up the precursor of the fibrinogen which, as we have seen, appears to exist dissolved in

[1] 'On a New Constituent,' &c., *ante.*

plasma. I do not know what this special substance is ; it is derived from the body separable by cold and also from leucocytes. I will remark that it is not contained in ordinary serum, and hence is not paraglobulin.

Peptone-plasma behaves in a totally different manner on heating, according as it contains a considerable amount of the body separable by cold or not. If this body be present, the plasma remains, on heating to 56–57°, perfectly clear for a short time—five to fifteen minutes ; it then becomes gradually turbid, and finally a dense flocculent precipitate forms ; but it is a very long time—hours—before this precipitate reaches its maximum.[1] If it be absent, the plasma does not give on prolonged heating to 56–57° any coagulum, and remains perfectly free from any precipitate till a very high temperature (80–90°) is reached. It becomes opalescent at high temperatures. In either case, whether the body separable by cold be absent or present, the plasma, if mixed with an equal quantity of 10 per cent. salt solution, remains free from any precipitate till 80–90°, or higher, is reached. The upper limit varies somewhat, often being as high as 95°.

It must be understood that the precipitate mentioned above as occurring in plasma at 56–57° may or may not contain the body separable by cold, but it appears in very much larger quantity there than does the latter, and represents the whole of the coagulable matter of the plasma. It would, therefore, appear that in the process of heating, the same special body is liberated from the substance separable by cold as is the case when the latter is acted on by carbonic acid—i.e. a body capable of converting the precursor of fibrinogen into fibrinogen precipitable at 56°, and coagulable with fibrin ferment. The presence of 5 per cent. NaCl prevents its development by means of carbonic acid, and also by means of heat.

In order not to obscure the point I have just been discussing, I have left out of consideration the very small trace of true fibrinogen pre-existent in peptone-plasma. This was mentioned at the beginning of the paper ; it always separates on heating the plasma to 56°, whether it has been cooled or not, or whether salt be present or not, but it is generally so very small in

[1] This slow clotting at 56–57° was observed by Fano.

quantity that it is difficult to see the minute flocculi that separate at 56°; hence the statements I made above as to the plasma remaining clear are practically correct. It must not, however, be forgotten that in some rare cases peptone-plasma does give a considerable clot with serum or ferment alone, and such a plasma gives a dense clot at 56°, whether the substance removable by cold be present or not.

Now it is a well-known fact that the injection into the veins of a strong solution of fibrin ferment, prepared in the ordinary way, is very rarely followed by any serious thrombosis. The same is true of defibrinated blood, and this is quite in conformity with the results I have described above.

This note is only a slight addition to the results I have already obtained in the coagulation question, and I am actively engaged in pursuing the subject.

I wish to take this opportunity of making a few statements with regard to the action of lecithin in producing coagulation. I have already shown that lecithin is an important factor in the coagulation of the blood. Since these publications,[1] my observations have only tended to completely confirm the statements I have already made; and I have, in addition, found that lecithin from the most varied sources, and lecithin prepared from the platinum salt, is perfectly active. The sources which have yielded an active lecithin are lymph-glands, blood, testis, brain, yeast. Moreover, I have found that it exerts its influence on other varieties of plasma besides those which I have already quoted, viz. peptone-plasma and cooled plasma. As to its exact mode of action I am not yet certain. It, no doubt, gives rise to the appearance of a large quantity of fibrin ferment, as was described in my note 'On the Origin of the Fibrin Ferment,' but I have reason to think that it has a further action.

The majority of the experiments on which the above paper has been founded were carried out at the laboratory of the Brown Institution.

---

[1] 'Further Observations on the Coagulation of the Blood,' *Journal of Physiol.* 1883; 'On the Coagulation of the Blood,' *Journal of Physiol.* 1884; 'On the Origin of the Fibrin Ferment,' *Proc. Roy. Soc.* 1884. [Collected Papers,' pp. 100, 113, 119.]

ON INTRAVASCULAR CLOTTING [1]

I PROPOSE to give a short description of a substance the introduction of which into the circulation of an animal causes almost instantaneous death from the complete fibrinous coagulation of the blood that it produces. This substance is a mixture or perhaps a compound of proteid and lecithin. It gives all the reactions of proteid and contains a large amount of lecithin. It may be prepared in large quantities from the testis and thymus of young animals, especially of calves, the method of procedure being to mince the organ, and after covering with water, to allow it to stand for a few hours, when it is centrifugalised until no further deposit is produced. The liquid is then rendered strongly acid with acetic acid, whereupon a bulky flocculent precipitate appears. This is collected by means of the centrifuge and thoroughly washed with water.

This precipitate dissolves readily in very dilute sodium carbonate. Injection of this solution into the jugular vein of an animal (such as a dog, cat, or rabbit) is followed by immediate death, thrombi being found at the post-mortem examination throughout the entire vascular system. But in order to obtain this result a certain quantity of the substance must be employed, since the effect is in proportion to the amount injected. About 1·5 grm. is enough to kill a dog of moderate size. If only a small quantity be injected, the thrombosis will be limited, and the blood drawn off from the carotid after the operation will not coagulate spontaneously. In this case it will remain fluid for about the same length of time as it does after the injection of peptone. It clots, however, rapidly if more of the substance, or if lecithin only, be added.

---

[1] [Translated from *Du Bois' Archiv*, 1886, p. 397. (A shorter account of this discovery was also published in the *Proc. Roy. Soc.* Feb. 4, 1886.—Ed.).]

I regard the active agent in the acetic acid precipitate as a compound of lecithin and proteid, and will now give the grounds for this view.

1. The precipitate has no effect whatever upon dilute magnesium sulphate plasma; it does not, therefore, contain any fibrin ferment. The blood withdrawn after the injection also contains only a mere trace of fibrin ferment.

2. The precipitate, when recently obtained, is soluble in dilute hydrochloric acid; if again precipitated by neutralisation, it retains its characteristic properties. But if a little pepsin be added to the hydrochloric acid solution, and the mixture be left to stand for a short time at 37°, a flocculent precipitate makes its appearance. If the products of digestion be neutralised and injected into the animal, no effect is produced. I may remark that this result cannot be due to the presence of the pepsin and peptone, for if some fresh precipitate be mixed with the products of digestion, and the mixture injected, death will ensue with the usual symptoms. The pepsin digestion gives rise to a decomposition which renders the precipitate ineffectual.

3. The larger portion of the lecithin may be removed from the precipitate by treatment with alcohol and extraction with ether. The residue is still soluble in dilute alkalies. But this solution, even if injected in very large quantities, will not produce any intravascular clotting, although the blood which is afterwards drawn off coagulates rather more slowly, and at times very slowly indeed.

A subsequent injection of lecithin alone has no apparent result.

With regard to the distribution of this substance, I may state that it is present in the juice of lymphatic glands. If these are minced, and squeezed out with 0·6 per cent. salt solution, and the liquid subjected to the action of the centrifuge, the leucocytes will be thrown down; and injection of the remaining fluid, entirely free from all cellular element, causes widespread intravascular clotting.

The substance above described may be precipitated from this liquid by means of acetic acid. The injection of the washed leucocytes has no effect. This observation was made on a former occasion by the author, but in view of the statements made by Groth, it has been repeated.

The stromata of the red blood-corpuscles of mammalian blood contain a similar substance and have the same action.

I devised a method, the description of which will be found elsewhere,[1] for isolating the stromata—i.e. for separating them from the constituents of the serum and from white corpuscles. When dissolved in Na_2CO_3 solution and injected into rabbits, these stromata have invariably produced fatal results, with extensive thrombosis; whereas the injection of a strong solution of hæmoglobin had no action. The well-known fact that the injection of laky blood often brings about intravascular clotting has hitherto been ascribed to the hæmoglobin.

The stromata contain a quantity of lecithin, besides roteids, and they show the same reactions towards hydrochloric acid and pepsin as the acetic acid precipitate does.

This paper is merely of the nature of a preliminary communication, in anticipation of more detailed accounts of my investigations on the coagulation of the blood.

[1] 'Zur Chemie der Blutkörperchen,' *Du Bois' Archiv*, 1881. [Collected Papers, p. 69.]

CLASS IV

THE CROONIAN LECTURE
ON THE COAGULATION OF THE BLOOD

THE CROONIAN LECTURE [1]

ON THE COAGULATION OF THE BLOOD

I. *The Relation of the Formed Elements of the Blood to Coagulation*

IT is generally accepted, at the present time, that the intervention of formed elements is necessary for coagulation, that the blood-plasma is lacking in certain of the constituents necessary for coagulation, and that these factors are supplied either by the white blood-corpuscles or by certain special corpuscles, to which the name of ' Blutplättchen ' is generally given. My observations have led me to an opposite conclusion. I find that the plasma contains, in solution, everything necessary for coagulation. And although I will not go so far as to affirm that the formed elements cannot exert any influence, yet, as I hope to show, it is very doubtful whether they play any part at all, and it is certain that they are not necessary. The proof of this is as follows. By injecting peptone into the blood of a dog, the blood drawn off directly after the injection is prevented from clotting. For the present we may leave out of the question how it is that peptone produces this effect. The result is always constant, and we can obtain from the blood drawn off, by subjecting it to the action of a centrifugal machine, a perfectly clear plasma free from all formed elements. This plasma, if left to itself, will remain fluid for a very long period—in fact, until putrefactive changes set in—but it will clot on subjection to very simple influences, influences which cannot be looked upon as being in themselves fibrin factors, but which must be regarded as simply . neutralising the restraining power exerted by the peptone injection on coagulation. Thus the plasma will clot when a stream

---

[1] Delivered April 8, 1886.

of carbonic acid gas is passed through it, or when diluted with water, or more slowly when diluted with ½ per cent. solution of common salt; further, it will clot imperfectly if filtered through a clay cell. Peptone-plasma[1] is entirely free from fibrin ferment, but in the process of coagulation fibrin ferment makes its appearance. By receiving blood direct from the blood-vessels into solutions of certain neutral salts the blood can also be prevented from clotting; and, in the same way as in the case of peptone-plasma, a plasma can be obtained entirely free from corpuscles. The plasma will be found to have different characters according to the solution which has been employed to prevent coagulation. Thus, if we receive into a 10 per cent. solution of common salt an equal volume of blood, we shall obtain a plasma which clots readily on dilution with four or five times its bulk of water. It will be convenient for me to allude to this plasma as 'sodium chloride plasma.' This plasma, before dilution, is quite free from fibrin ferment; after clotting has taken place ferment is found to be present. If, instead of using a 10 per cent. solution of sodium chloride, we use a solution of sulphate of magnesia, we shall get a plasma which does not clot or form ferment on dilution. In order to make this plasma clot, ferment must be added. This variety will be referred to as 'magnesium sulphate plasma.' It is obtained most conveniently, in the case of the dog, by taking ½ vol. of saturated solution of sulphate of magnesia to 1 vol. of blood. Another plasma can be got from the blood of the horse by cooling it to 0° and allowing the corpuscles to settle; by filtering it, still at a low temperature, through several folds of thick filter-paper the plasma becomes almost entirely free from corpuscular elements, and has but very little power of clotting by itself. Apparently then, these two last kinds of plasma afford no support to the statement made concerning the spontaneous coagulability of the plasma. But the result is explained by the following fact :—

The body which has the power of initiating coagulation is readily separable from the plasma by cooling or by a certain strength of magnesium sulphate. This body is a new substance

[1] Peptone-plasma—the plasma from peptonised blood.

I have discovered; it forms such an important element in the question of coagulation that without a knowledge of its existence the most experienced observer could hardly fail to fall into error. It has been mentioned that peptone-plasma clots when a stream of carbonic acid gas is passed through it, and that at the same time fibrin ferment is formed. If the plasma be placed in ice it rapidly becomes turbid, and in course of time a considerable precipitate forms. This is the new substance. If the cooling is prolonged until all this substance disappears, the plasma will be found to have lost its power of clotting with CO_2 and of giving rise to fibrin ferment. If only part of the new substance is removed, the plasma coagulates more slowly and less ferment is formed; and by again adding some of this body to the plasma which has been deprived of it, the power of spontaneous coagulability is restored to the plasma. If peptone-plasma is mixed with an equal volume of 10 per cent. NaCl solution, it retains its power of clotting on dilution, and cooling does not give rise to a precipitate. In fact, peptone-plasma, treated in this way, behaves just as normal NaCl plasma does; it retains its power of clotting on dilution for an almost unlimited period. If, however, about $\frac{1}{5}$ of its bulk of a saturated solution of magnesium sulphate is put to peptone-plasma, it is at first capable of spontaneous coagulation; but on standing—for a short time at ordinary temperature, and more quickly at a low temperature—there arises in the plasma a precipitate, upon removal of which the plasma will not clot on dilution; it completely resembles the normal $MgSO_4$ plasma. It may now be understood how it is that this form of plasma— i.e. normal $MgSO_4$,—presents an exception to the general statement brought forward.

The 'cooled' plasma must be considered in somewhat greater detail, for the doctrine of the participation of the white corpuscles in coagulation has been founded in great measure upon experiments made with this variety of plasma. The researches of Alexander Schmidt are well known, but it may be advisable to quote some of the more essential observations bearing on this subject. The blood of the horse is used, and is prevented from clotting by collection in vessels surrounded with ice. The blood is thus maintained at a low temperature until the corpuscles sink, the red falling much more readily than the white. After a time, in spite of the cold, coagulation sets in, beginning first and most extensively in the lower layers of the colourless plasma, above the sediment of red corpuscles, where there are most white corpuscles; it is scanty and later in the upper layers of the plasma, where there are but few corpuscles. Now it must be remembered that this new substance separates from the plasma when cooled. Schmidt did not know of the existence of this substance, nor could he, from his method of experimentation, obtain any definite evidence of such a substance. But it will be obvious that the greater coagulability of the lower layers of the plasma may be due, not to the greater number of white corpuscles present, but to this new substance being precipitated out by the cold and sinking to the lower layers of the plasma. Indeed, Schmidt himself mentions that in these lower layers of plasma not only are great numbers of white corpuscles to be seen, but also a large quantity of granular matter. He discusses this granular matter, and comes to the conclusion that it is the *débris* of white corpuscles which have broken up in spite of the cold. But he has no conclusive evidence to offer in support of this view. I think that my experiments leave no doubt as to what this granular precipitate is—*i.e.* this new body separable by cold from the plasma; and inasmuch as we have seen, from the account of my observations above given, that this substance is of great importance in initiating coagulation, these experiments of Schmidt's lose their force as conclusive proof of the part taken by the white blood-corpuscles in coagulation. The same objection can be urged against the experiments of Schmidt on the cooled plasma filtered through thick filter-paper. This plasma clots very slowly and forms but little ferment. But these results cannot be attributed to the absence of corpuscles, since, by the process of cooling, the new body has been removed as well as the corpuscles. So far, then, the participation of the white corpuscles is an open question, but with lymph-corpuscles the case is different.

In a paper read before the Royal Society in 1881,[1] I showed that isolated lymph-corpuscles had the power of inducing coagulation in a marked degree. This was the first definite proof that any form of corpuscle could contribute to the process of coagulation. My experiments were made with peptone-plasma, and as they have been repeated with the same result by other observers on different kinds of plasma, there is no doubt as to their correctness. If lymph-corpuscles are added to peptone-plasma, or to cooled plasma, clotting rapidly takes place. This might be regarded as strong evidence in favour of the white corpuscles of the blood also being agents in the process of coagulation, and at the time these experiments were made I did so regard it. But it is only valid if there is adequate proof of the identity of white blood-corpuscles and lymph-corpuscles. This supposition is generally made ; but some observations have led me to doubt whether it is a true one, and have also thrown grave doubts on the capability of the white blood-corpuscles to initiate coagulation. The experiments are as follows :—A dog having been injected with peptone, a small quantity of its blood was drawn off, which remained fluid, but clotted immediately on the addition of a small quantity of lymph-corpuscles. A great number of lymph-corpuscles suspended in salt solution were now injected into the jugular vein of the dog. This had no effect on the animal. Some minutes later another portion of blood was drawn off, which, like the first, remained fluid. A fresh addition of lymph-corpuscles to this second portion of blood was immediately followed by coagulation. In a second experiment the conditions were the same, but instead of injecting leucocytes into the circulation generally, they were injected into an isolated vein, filled of course with blood. Here again the blood drawn off from this vein after the injection of the lymph-corpuscles still remained fluid, in spite of the fact that it contained very large quantities of leucocytes.

I have made a similar observation on a dog not treated with

[1] 'On the Relation of the White Blood-corpuscles to Coagulation,' *Proc Roy. Soc.* 1881 [This paper is almost identical with Part III. of 'The Chemistry of the Blood-corpuscles,' *vide* Collected Papers, p. 89.—Ed.]

L

peptone.   A very large quantity of leucocytes was injected into
the circulation.   The blood drawn off clotted very slowly, taking
when left to itself half an hour.   This blood was swarming
with leucocytes, and yet the addition of a few more immediately
set up clotting.   The leucocytes, when they have got into the
living blood-stream—*i.e.* the blood in the vessels—seem to lose
the power they previously possessed of setting up coagulation.
This must be regarded as weighty evidence against the participation of the white corpuscles ; but still more may be adduced
against it.   Lymph-corpuscles,[1]   when   treated   with   strong
solutions of neutral salts, undergo a very peculiar change, becoming converted into a strong mucoid material.   If a small
quantity of leucocytes are put into blood, the addition of some
strong salt solution converts the whole into a viscid slimy mass ;
but the addition of salt solution to blood alone does not effect
this change.

II. *The Coagulable Matters of the Plasma*

By coagulable matters are meant the substances which yield
fibrin.   The plasma contains all the precursors necessary, and
we now come to the consideration of these bodies.   It has long
been known that a substance can be obtained from the plasma
which, though it is not fibrin, can readily be converted into this
substance by appropriate means.   This substance is called
'fibrinogen.'   Though first brought into notice by Alexander
Schmidt, it was much more accurately studied by Hammarsten,
and there is no doubt that such a substance can be obtained
from the blood-plasma.   It is further certain that this substance
can be converted into fibrin by means of the fibrin ferment.
This has reference to the substance Hammarsten calls 'fibrinogen.'
According to Schmidt, fibrinogen requires paraglobulin and
ferment to convert it into fibrin.   Now the first point to which
prominence must be given is that the plasma does not contain
any body which will clot either with ferment or with paraglobulin, or contains at most an infinitesimal trace of such a body.
It is true that fibrinogen can be obtained from the plasma, but

[1] Isolated corpuscles of lymphatic glands.

it acquires its characteristic properties as a result of the means used to isolate it. The fibrin-yielding matter, as it exists in the plasma, is quite different from the fibrinogen of Hammarsten or Schmidt. It can, however, easily be converted into this substance, and is so converted by the means usually adopted for its isolation. Again, the fibrin-yielding matter, as it exists in the plasma, is not a single body, but a mixture of two very closely allied substances. To meet this state of things, it is necessary for me to adopt special names for these bodies. Inasmuch as they are all convertible into fibrin, and are nearly related to one another, I shall for the present designate them as A, B, and C fibrinogen. The chief reason I have for adopting this somewhat clumsy nomenclature is the fact that, though they present well-marked differences from one another, they are convertible one into another; and in this process intermediate bodies occur which have some of the properties of the one, and some of the properties of the other, variety of fibrinogen. This will become more apparent as we proceed, but I use these names more with the object of facilitating description than with the view of establishing them permanently. As will be seen, both kinds of coagulable matter are very readily changed; hence some of their properties can only be ascertained indirectly by the behaviour of the plasma which contains them.

A-fibrinogen is the name given to the substance which separates from the plasma on cooling or on adding a certain amount of sulphate of magnesia (*vide* p. 142). It is the body which endows the plasma with spontaneous coagulability; it is the body which gives rise to the ferment in the plasma. It was first found that plasma free from any formed elements, and free from any trace of fibrin ferment, clots and yields fibrin ferment without the addition of any fibrin factor. It was then found that with the removal of *A-fibrinogen* the plasma lost its power of spontaneous coagulability and its power of spontaneously forming fibrin ferment. *A-fibrinogen* separates from peptone-plasma by cooling the latter; a very slight lowering of temperature suffices to produce a precipitation of a certain amount of the substance. Prolonged cooling down to 0° removes either the whole or nearly the whole of the substance. When first separated, it presents a very peculiar appearance nder the microscope. Instead of looking like a collection of irregular granules, as proteids generally do, the precipitate is found to consist entirely of small rounded homogeneous discs [1] of nearly equal size; but when they have been separated for a certain time, they run together into granular lumps or threads. The appearance of these discs is absolutely identical with the Blut-plättchen of Bizzozero, the special formed elements above alluded to; and there can be little doubt but that the bulk of the bodies which are looked upon as Blutplättchen are not formed elements at all, but the precipitated *A-fibrinogen*. When first separated it is readily redissolved on warming the plasma, but when it has become collected into lumps and threads it is no longer soluble on warming. The same is true as regards its solubility in 4 per cent. NaCl solution. It must, therefore, undergo a change when it has been separated for some time. The interval which must elapse before this change is accomplished varies very considerably in different specimens of plasma. By the time it has been collected on the centrifuge, it has generally undergone marked changes in its solubilities, the degree to which these changes take place varying in different forms of plasma and with the time occupied in its collection; it now greatly resembles a disc of fibrin, but it is of a peculiar slimy character. The sliminess, however, disappears when it is pressed together between the fingers or filtering-paper; and it is then quite indistin-guishable from ordinary fibrin, so far as appearance and elasticity are concerned. Examined chemically, it is found to consist in the main of proteid, with a considerable quantity of lecithin. It is very little affected by solutions of neutral salts; it swells up somewhat, but dissolves, if at all, extremely slowly. Dilute alkali has a similar action upon it. In dilute hydrochloric acid it mostly dissolves, leaving only a swollen-up shell, as it were. The first effect of the acid appears to be that of making it more opaque, and this result is more marked in the case of acetic

---

[1] These discs fuse together under certain conditions, and give rise to larger discs, which have a great resemblance to blood-corpuscles.

acid. But its behaviour towards hydrochloric acid varies according to circumstances : if it has been separated from a well-peptonised plasma, and for a limited period only, it behaves as above stated; but if from a plasma less fully peptonised, it is quite indistinguishable from ordinary fibrin. *A-fibrinogen*, as it exists in the plasma, is not acted on by the fibrin ferment, for if we add ferment to peptone-plasma, and allow a prolonged period for the ferment to exert its influence, the plasma still gives the precipitate on cooling with just the same readiness. Further, as it exists in the plasma, it does not clot with heat till a high temperature (over 80° C.) be attained. If the plasma be made to clot either by CO_2 or by adding leucocytes, the *A-fibrinogen* disappears in the clot formed. *A-fibrinogen* is apparently dissolved chiefly by the alkali of the plasma. Neutralisation greatly favours its precipitation by cold; a slight increase of alkalinity prevents separation. It is precipitated from the plasma by cold, by certain strengths of sulphate of magnesia, or by acidifying the plasma with dilute acetic or sulphuric acid. When first separated it is readily redissolved either by neutral salt solutions or dilute alkali; if it has been separated for a longer time it loses its solubility and appears as a fibrinous substance.

In the first section the fact was mentioned that the addition of magnesium sulphate causes a precipitate, the removal of which takes away the power of spontaneous coagulability. The experiment should be performed by adding just so much saturated solution of $MgSO_4$ as is insufficient to produce an immediate precipitate. On standing for a time a precipitate arises, which, when collected, is found to have just the same properties as the cold precipitate. The temperature of coagulation by heat of *A-fibrinogen* is high—over 80°.[1]

B-fibrinogen forms the bulk of the coagulable matter of the plasma. It does not precipitate out on cooling the plasma, but is precipitated by neutral salts or by acids, in the process of which it undergoes changes. As it exists in the plasma, it

[1] 'On the Fibrin-yielding Constituents of the Blood Plasma,' *Proc. Roy. Soc.* 1885. [Collected Papers, p. 129.]

possesses the following properties : it does not clot with fibrin ferment; it does clot with lymph-corpuscles. The temperature of heat-coagulation is high—over 80° C.[1] Chemically it consists of proteid and lecithin.

Both *A*- and *B-fibrinogen* are insoluble in a slight excess of dilute acetic or sulphuric acid, thus presenting a marked contrast to paraglobulin, which is readily soluble in dilute acid.

C-fibrinogen is the fibrinogen of Hammarsten; it is either absent from the plasma or present in very small quantities. Its characteristic features are well known for distinctive purposes. It clots readily with fibrin ferment, it coagulates on heating at 55° C.; it does not clot with leucocytes. Chemically it consists of proteid and lecithin.

The relation between the coagulable matters of the plasma and the fibrinogen of Hammarsten will be more easily understood from the following. We take peptone-plasma[2] (coagulable with leucocytes but not with ferment) and precipitate it by adding an equal quantity of salt solution. The precipitate redissolves in dilute salt solution, and can again be precipitated by adding salt in substance. But the first precipitate—*i.e.* the precipitate as it comes from the plasma—is very different from the second. The first precipitate is finely flocculent or granular ; the second is coarsely flocculent, and obviously resembles fibrin in appearance. The first precipitate is much more readily soluble in dilute salt solutions than the second. The solution of the first precipitate clots with fibrin ferment slowly, but it also clots readily with leucocytes. The solution of the second precipitate clots with ferment but not with leucocytes.

We have then a plasma not coagulable with ferment, but coagulable with leucocytes ; a first precipitate from this plasma coagulable with both these, and a second precipitate which, when dissolved, is coagulable only with ferment. From these experiments we may conclude that the coagulable matter as it exists in the plasma is a substance having widely different characters

[1] *Loc. cit.*

[2] Magnesium sulphate plasma from horses' blood is also particularly suited for this purpose.

from the coagulable matter isolated from the plasma; or, in other words, by the processes of precipitation and isolation the *B-fibrinogen* of the plasma is gradually converted into the classical (*C*) fibrinogen. The first precipitate is, however, evidently an intermediate product, and in certain forms of plasma the coagulable matter appears to be chiefly in the form of this intermediate product. Thus cooled plasma clots slowly with ferment, weakly peptonised plasma does the same, but they both coagulate much better with leucocytes. Strong dilution of the plasma is sufficient to completely alter its properties. Thus from strong magnesium sulphate plasma a precipitate is obtained which, dissolved in a little dilute salt solution, clots readily with leucocytes. If the solution of the precipitate, or the plasma itself, be much diluted with water, we get a solution which does not coagulate with leucocytes, but only with ferment.

In describing the above I have not considered whether the plasma contains both *A*- and *B-fibrinogen* or only *B*. The end product is the same in either case—*i.e.* a solution is obtained agreeing in all respects with Hammarsten's fibrinogen. But if the plasma contains *A-fibrinogen*, the first precipitate is spontaneously coagulable on solution; if the solution be rapidly precipitated again before clotting occurs, we get *C-fibrinogen* as the end product.

The Plasmine of Denis.—By saturating sodium sulphate plasma with solid sodium sulphate, Denis obtained a precipitate which, when redissolved in dilute salt solution, clotted spontaneously. The result of subsequent investigations tended to show that this was a mixture of two proteids, one of which was called 'fibrinogen,' the other 'paraglobulin,' with these being fibrin ferment. Fibrin was supposed to arise from the interaction of these three bodies (Schmidt), or to be due merely to the action of the ferment on one of them—the fibrinogen.

The Plasmine of Wooldridge.—No doubt the plasmine of Denis may contain paraglobulin, but we will now describe a method by which a 'plasmine'—*i.e.* a spontaneously coagulable precipitate—can be obtained, from which the presence of paraglobulin is absolutely excluded. It is obtained by treating peptone-plasma

which has not been cooled, with very dilute sulphuric acid (6 c.c. of
the strong acid to one litre of water). This is added to the plasma
till it is markedly acid in reaction. The precipitate does not
arise on neutralisation when the reaction is only just acid;
only when there is a strong acid reaction does the precipitate
appear. It will be remembered that, although paraglobulin may
be precipitated on neutralisation under certain conditions, it is
extremely soluble in the slightest excess of mineral acid. In
fact, on isolating paraglobulin from serum by the method·of dilution and CO_2, I have always found that it is very readily soluble
in the slightest trace of the dilute sulphuric acid above mentioned; we may hence safely conclude that the precipitate which
arises in peptone-plasma when the latter has a strongly acid
reaction does not contain any paraglobulin. Nevertheless, the
precipitate obtained from peptone-plasma by dilute acid is a
plasmine, for if collected and washed and redissolved in dilute
alkali (it dissolves whilst the reaction is still faintly acid) it clots
spontaneously, yielding a dense fibrin clot. But this is only the
case if the plasma contains *A-fibrinogen*. If the latter has been
removed by cooling, the precipitate is not spontaneously coagulable; it clots only on addition of fibrin factors. It must be
noted that the plasma and the precipitate are both free from
fibrin ferment. Two additional points must be referred to with
regard to this acid precipitate. In the first place, the *A-fibrinogen* is more readily precipitated by the acid than the *B-fibrinogen*, for we can obtain by fractional precipitation a first precipitate containing *A*- and some *B-fibrinogen*, which is spontaneously
coagulable, and by further addition of acid a second precipitate
of *B-fibrinogen* only, not spontaneously coagulable, but coagulable on addition of ferment, &c. Secondly, as above mentioned, *A-fibrinogen* very easily undergoes a fibrinous change, so
as to become insoluble in dilute alkali. Whenever this occurs,
the solution of the acid precipitate is not spontaneously coagulable; it only clots when additions are made. In dissolving the
precipitate a small part remains behind as a fibrinous substance,
having the same character as the changed *A-fibrinogen* has when
simply separated by cold. If acid is added to the plasma until

no further precipitation arises, even on standing, the plasma has entirely lost its power of coagulating.

It is quite true that plasmine consists of a mixture of two bodies, but both these bodies are sharply to be distinguished from paraglobulin by their insolubility in dilute sulphuric or acetic acid. A plasmine, a spontaneously coagulable precipitate, can be obtained from (1) peptone-plasma which has not been cooled, (2) salt plasma which does not contain too much salt —e.g. 4 per cent. salt solution. Cooling, or a slight excess of salt, removes one of the necessary factors—i.e. A-fibrinogen, a body which has not the slightest resemblance to paraglobulin.

III. *The Chemical Processes of Coagulation*

In considering the question of the chemical processes of co-agulation, it will be convenient to divide the subject into three parts. The first will contain evidence that lecithin is a factor, and that the interaction between the *A-* and *B-fibrinogen* is of the nature already described. The second will treat of a substance which causes intravascular clotting, the mode of action of which bears out the views above expressed. The third will be devoted to the consideration of the relations of the fibrin ferment.

Allusion has already been made to the fact that peptone-plasma is coagulable by passing through it a stream of carbonic acid gas. In cooling the plasma, *A-fibrinogen* disappears, in consequence of which the plasma loses its power of clotting with CO_2. The conclusions drawn from this experiment are that *A-* and *B-fibrinogen* in peptone-plasma are inhibited in their interaction, which should normally occur when the blood leaves the vessels, by the peptone injection. Neutralising the plasma by CO_2 (or other acid) overcomes this inhibition, and the interaction takes place, resulting in the appearance of fibrin. In what way the peptone injection effects this inhibition is still not quite clear; but, bearing in mind the rapid disappearance of injected peptone, it is not improbable that it unites either with the *A-* or *B-fibrinogen*, or with both, thus preventing their action until some 'disturbance' is set up in the plasma. As in the case of salt-plasma, the presence of strong salt solution prevents the interaction, while dilution removes the inhibitory influence and enables the interaction to take place.

It is not necessary to replace *A-fibrinogen* itself; it is only necessary to add one of its constituents—lecithin—in order to restore the power of spontaneous coagulability. To illustrate this it will be advantageous to describe an experiment. We take some peptone-plasma, and find that it clots readily within some ten or fifteen minutes with CO_2, the plasma being then put in ice for two days and the precipitate removed. It does not now coagulate with CO_2, or at most gives a minute trace in twenty-four hours. But if lecithin be added to this plasma and CO_2 passed through, it clots with rapidity in from five to ten minutes.

When plasma clots owing to the presence of *A-fibrinogen*, fibrin ferment is formed; when it coagulates owing to the presence of lecithin, not only fibrin, but also fibrin ferment make their appearance.[1] There is thus a marked parallelism between the action of *A-fibrinogen* and that of lecithin in inducing coagulation, and since *A-fibrinogen* contains lecithin, it is not unreasonable to suppose that it is by the agency of this latter body that the *A-fibrinogen* is enabled to induce the coagulation of *B-fibrinogen*.[2] But is there any evidence that the addition of lecithin to *B-fibrinogen* converts this into fibrin? The above experiment is not conclusive on this point, owing to the fact that fibrin ferment makes its appearance; and though fibrin ferment has no action on *B-fibrinogen* in its typical condition, the latter is very readily altered so as to be acted on by the ferment (*vide supra*), and as a matter of fact, in the above experiment, fibrin ferment is able to replace lecithin to a certain extent, although coagulation takes place much more rapidly with the latter substance than with the ferment. By a

[1] To avoid confusion, it may be stated that neither lecithin nor any of its products of decomposition is identical with fibrin ferment.

[2] With regard to ferment formation, *vide* 'On the Origin of the Fibrin Ferment,' *Proc. Roy. Soc.* 1883 [Collected Papers, p. 119]; and 'On a New Constituent of the Blood' &c., *Proc. Roy. Soc.* 1885 [Collected Papers p. 124.]

slight modification of the above experiment this difficulty is overcome, and a clearer insight into the process of coagulation is also attained. For this experiment we must take a plasma from a dog less completely under the influence of peptone than it was for the former experiment. If lecithin only be added to this plasma, complete clotting will rapidly take place, depending in a remarkable way on the temperature. If some of the plasma be put into ice, the precipitation of *A-fibrinogen* takes place in the usual manner, but, as is always the case in moderately peptonised plasmata, it changes rapidly into the fibrinous substance. The amount of fibrin is, of course, very small, but if after the removal of this we add lecithin to the plasma, and again place it in ice, complete coagulation of the plasma takes place. The mode in which this occurs is extremely interesting. It always commences on the top, where the fluid is coldest, and travels gradually downwards, so that the upper half may be completely solid whilst the lower is still quite fluid. Again, coagulation first appears in the form of little irregular lumps round the edges of the vessel and flowing freely in the fluid, the lumps gradually coalescing into a compact mass. The fibrin thus formed is ordinary fibrin. When clotting takes place in this way, fibrin ferment does not always make its appearance,[1] but it is certain that even when present, the formation of fibrin cannot be referred to it, for if very strong ferment be added and the plasma kept in ice, no clotting takes place. We may safely conclude, therefore, that the appearance of fibrin is caused by the direct addition of lecithin to *B-fibrinogen*. As before stated, temperature has a great influence on the process of coagulation in this instance. If, after adding lecithin to the plasma, it is kept at 37° C., not a trace of clotting will be found ; if kept at the ordinary temperature, imperfect clotting occurs.

Normal ' cooled ' plasma behaves just as does weakly peptonised plasma towards lecithin ; that is, it clots rapidly at a temperature of about 0°.[2] Salt-plasma does not clot with

---

[1] Whether it does or not appears to depend on the rapidity of the process.
[2] *Journal of Physiology*, 1884. [Collected Papers, p. 113.]

lecithin, and it can be shown that the presence of the strong salt solution prevents its action. The first precipitate from salt-plasma, if dissolved in a trace of alkali, does clot with lecithin. (The first precipitate is, as I have explained above, more or less modified *B-fibrinogen*, or rather, a body intermediate between *B-* and *C-fibrinogen*.) This experiment with lecithin is best done with the blood of the horse, as in the case of dog's blood the change of the fibrinogen existing in the blood to *C-fibrinogen* often takes place with extreme rapidity; the first precipitate may rapidly become more or less insoluble, and the solution may contain nothing but *C-fibrinogen*, which clots only with fibrin ferment. The horse's blood is received direct into strong $MgSO_4$ solution (1–3 or 4); to the plasma obtained from this blood NaCl in substance is added (up to 16 per cent.) The precipitate, collected by the centrifuge, is firmly pressed between folds of filtering-paper and dissolved in very weakly alkaline water : it clots on the addition of lecithin. I have performed this experiment several times with great care and quite successfully, though not invariably so. The difficulties of the experiment are numerous, and its success depends on the rapidity with which the precipitate is collected, on the amount of fluid and degree of alkalinity of the fluid used, on the amount of salt present, on the greater or less completeness of the lecithin emulsion.

Intravascular Clotting.—A theory of coagulation which does not tell us how to induce coagulation of the blood in the vessels must be very imperfect ; and it is at this point that the previous theories—the ferment theory and the ferment and paraglobulin theory—break down ; for it is beyond all question that the injection of serum, containing both ferment and paraglobulin, is not followed by intravascular clotting. The observations I have been able to make are in complete accordance with the lecithin theory. There exists a substance[1] the injection of a solution of which into the blood causes the most widespread intravascular clotting ; and, as will be shown, the action of this

---

[1] 'On Intravascular Clotting,' *Proc. Roy. Soc.* 1886 and *Du Bois' Archiv*, 1886. [Collected Papers, p. 135.]

substance depends on the lecithin it contains. I first obtained this substance from the testis and from the thymus gland ; subsequently I found it in the fluid of lymphatic glands. From this fact, and from the fact that it passes into fibrin, I have for the present designated it as ' lymph-fibrinogen.' In order to obtain it from lymphatic glands, all that is necessary is to mince these finely, to squeeze out with $\frac{1}{2}$ per cent. salt solution and to filter through a linen cloth, and then, by means of the centrifuge, to free the fluid of all corpuscular elements. The fluid thus obtained is rendered strongly acid with acetic acid, whereupon a dense precipitate arises. This is collected and washed with distilled water. This solution dissolves readily in dilute alkali, and when injected into the circulation of an animal causes instant death, owing to the most widespread intravascular clotting.

The chemical characters of this substance[1] are as follows : It consists of proteid with a large amount of lecithin. In a perfectly fresh state it is insoluble in acetic acid, unless the latter be very concentrated. It swells up in water and forms an apparent solution, but not a true one, as it will not filter through a clay cell. It is readily soluble in dilute alkali. In dilute hydrochloric acid it is apparently soluble—i.e. soluble in the sense that the casein in milk is soluble. It is insoluble in dilute sulphuric acid (0·5 per cent.). The hydrochloric acid solution gives a peculiar reaction with pepsin. If pepsin be added and the solution maintained at 37° C., there forms, after a short time, a flocculent precipitate containing proteid

[1] The substance or substances producing intravascular clotting can be obtained from various sources—testis, thymus, lymph-glands, brain, stroma of the red corpuscles. They swell up greatly with water, so as to form an apparent solution, though if it be filtered through filter-paper very little goes through. They are soluble in neutral salt solutions and insoluble in dilute sulphuric or dilute acetic acid. They swell up to a greater or less extent in dilute hydrochloric acid, and leave behind a great residue on peptic digestion. They are precipitated by an excess of salt. They differ slightly among themselves, particularly as regards the apparent ' solubility ' in dilute HCl. They could, of course, be considered as ' globulins,' but they are as totally different from paraglobulin as can be conceived, and should, I think, be looked upon as a special group of ' fibrinogens.' The fibrinogens in the blood are closely allied to them. They all contain a large quantity of lecithin.

and lecithin, whilst peptone is found in solution. On standing the substance very soon loses its power of dissolving in
hydrochloric acid. It is soluble, though not very readily, in
solutions of neutral salts, though again, on standing, it speedily
loses this power. Its action on the blood is this : within a few
seconds of injection it produces intravascular clotting, the
extent of which depends on the quantity of substance employed.
A very small quantity may be injected without giving rise to
any ascertainable thrombosis, whilst a large amount will produce
a complete fibrinous clot of the whole vascular system. To
effect this in a rabbit, from $\frac{1}{2}$ to 1 grm. of the solid substance is
necessary; in a dog, from 1 to 2 grms., according to its size.
The fibrin which is formed in the vessels is not to be distinguished from ordinary fibrin. The blood which is drawn off
after injection of this substance deserves very careful attention.
It does not coagulate, if left to itself, for a very long time,
remaining as a rule fluid for at least twenty-four hours; finally,
however, it clots firmly. It clots at once if some of the solution
of lymph-fibrinogen be added to it. Hence it is evident that
this substance disappears when injected into the blood ; and, as
will appear shortly, it no doubt forms the greater part of the
intravascular clot. For, in spite of the very extensive intravascular clotting, the shed blood contains enough fibrin-generators to give a solid and complete coagulation. By the injection,
however, of a very large quantity of the lymph-fibrinogen, the
amount of fibrin-generators in the shed blood can be obviously
diminished or be even made to disappear entirely. This account
of the shed blood applies more particularly to the blood of the
dog. In the case of the rabbit, the injection of the substance
in any considerable quantity leaves the shed blood with only a
trace of fibrin-generators, or with none at all.

The shed blood clots very readily with lecithin, and it also
clots, though very slowly, with fibrin ferment, crop after crop
of fibrin being formed. Thus one specimen of plasma, with
some very strong ferment solution, took two and a half hours
for complete coagulation ; with lecithin, clotting was accomplished in less than three minutes. The shed blood itself

contains but a minute trace of fibrin ferment, even when the intravascular clotting has been of the most extended nature. The character of the plasma then, after the injection of lymph-fibrinogen, is similar to that of 'cooled' plasma. It has traces of ferment, and ultimately, if left to itself, it clots. It clots slowly on the addition of ferment ; it clots with lymph-corpuscles, with lecithin, and with lymph-fibrinogen. So far as my observations have gone, the clotting of the plasma cannot be accelerated by any simple means, such as dilution or neutralisation. It would, therefore, seem to contain only *B-fibrinogen* in a slightly modified form.

When clotting takes place in the extravascular blood by means of lymph-fibrinogen, much ferment is produced, whereas in the case of intravascular clotting but a trace of ferment is formed. I do not know on what this difference depends. It is to be noted that the intravascular fibrin is ordinary fibrin.

It is, of course, difficult to say that the proteid-lecithin mixture or compound, which I have designated as 'lymph-fibrinogen,' consists only of this and nothing more ; but I shall endeavour to show, as we proceed, that there is every reason for thinking this substance to be the active agent ; and I shall also point out that any disturbance between the lecithin and proteid of which it is composed destroys its power. In the first place, lymph-fibrinogen does not contain any fibrin ferment ; it is quite inoperative on diluted $MgSO_4$ plasma, or on a solution of Hammarsten's fibrinogen, both of which clot very readily with fibrin ferment. As already stated, the body undergoes apparent solution in water and in dilute HCl ; the filtrate through a clay cell, which is free from the proteid-lecithin compound, is entirely inactive so far as coagulation is concerned.

I have alluded to the peculiar dissociation which takes place with pepsin and hydrochloric acid. What occurs is that the lecithin, together with a little proteid, is separated from the bulk of the proteid, which becomes peptone. This interference with the substance annihilates its power of inducing coagulation. I think great weight should be laid on this experiment. Again, if the substance is treated first with alcohol and then with ether, the bulk of the lecithin is removed; it still retains its solubility in dilute alkali (though it is with somewhat more difficulty soluble), but it does not produce intravascular clotting.[1]

If all these statements be fully considered, I think the conclusions are justified—1st, that the active agent is this proteid-lecithin mixture or compound; and, 2nd, that the lecithin it contains is essential to its action.

Returning to the action of lymph-fibrinogen on extravascular blood, we have to observe that it causes rapid coagulation in peptone-plasma, and disappears in the clot formed—that is to say, the serum from the clot is free from this substance; at the same time, a great deal of ferment is formed. Neither plasma nor the substance have any ferment, but when clotting has occurred, a large amount of ferment is found in the serum. Lymph-fibrinogen will clot with the first precipitate from peptone-plasma, or with the first precipitate from salt-plasma; but if this precipitate is altered by a second precipitation, or simply by being dissolved in a large quantity of fluid, lymph-fibrinogen has no effect upon it. The same is true of diluted $MgSO_4$ plasma. Generally speaking, lymph-fibrinogen acts when lecithin acts.

In reference to its distribution, I may add that the stromata of the red blood-corpuscles [2] are made of a substance similar to, if not identical with, lymph-fibrinogen; and, like the latter, its solutions produce extensive intravascular clotting. The injection of laky blood has been long known to bring about this result, which had previously been referred to the hæmoglobin. There are strong reasons for thinking that lymph-fibrinogen is the precursor of the *A-fibrinogen* of the blood. It certainly exists in the fluid of the lymphatic glands, and hence must probably reach the blood. In its chemical reactions it closely resembles *A-fibrinogen*, and, like the latter, has the power of causing *B-fibrinogen* to clot and of producing fibrin ferment. The *A-fibrinogen* of the blood is largely increased by a fatty diet [3]—a result which might be explained by the fact that fat passes

---

[1] This has only been tried in the case of the dog.

[2] *Practitioner*, March 1, 1886. [Collected Papers, p. 167.]

[3] 'On a New Constituent,' &c., *Proc. Roy. Soc.* 1884. [Collected Papers, p. 124].

chiefly through the lymphatics of the intestine, and feeding on fatty food may thus increase the flow of lymph-fibrinogen into the blood.

I have already stated that the amount of the intravascular clotting caused by the injection of lymph-fibrinogen is proportional to the amount of the latter injected. If a very large quantity be injected, it will exhaust the whole of the fibrin-yielding matter of the blood, and the shed blood will not give any clot at all. In the rabbit this is easily attained, but in the dog, owing to the fact that a very rapid and complete thrombosis of the whole portal system, and consequent fatal fall of blood-pressure, take place, the shed blood, as a rule, contains more or less coagulable matter, as above described. Nevertheless, it may by large injections be reduced to very little or disappear altogether; and in all cases, when the shed blood is at all coagulable, it readily clots on the addition of more lymph-fibrinogen. These facts are to be specially noted in connection with what follows.

The Fibrin Ferment.—Of the chemical nature of this body we know nothing. It was formerly supposed that the fibrin ferment, which is absent from the circulating blood, took origin by a 'breaking up' of the white corpuscles, consequent on the death of the latter. It was only under such a supposition that a theory like that of Hammarsten was possible. Recent investigations have, however, thrown a different light on the mode of origin of the fibrin ferment.[1] In the first place, it is quite certain that the ferment can arise from the plasma alone, provided *A-fibrinogen* is present; secondly, the addition to plasma of various bodies which are not in themselves the ferment, and do not contain any ferment, will cause the plasma to clot, and at the same time will give rise to fibrin ferment. Lymph-corpuscles, lymph-fibrinogen, lecithin, will cause the plasma to clot and a large amount of fibrin to appear; and, as above explained, lecithin will produce clotting under conditions—such as cold—which prevent the action of the ferment. So far as I know, the presence of

---

[1] 'On the Origin of the Fibrin Ferment,' *Proc. Roy. Soc.* 1884. [Collected Papers, p. 119.]

lecithin is essential for the production of ferment; it is at least positive that the addition of various substances or tissue elements which are rich in lecithin to plasma will give rise to coagulation and to the appearance of much fibrin ferment, and that lecithin itself will produce a similar effect. The ferment must, then, arise as the result of a complex reaction. It is not a question of leucocytes dying and liberating ferment.

Hammarsten's fibrinogen clots readily with ferment or serum, but neither lecithin, lymph-corpuscles, nor lymph-fibrinogen have any action on it. The bodies which, when added to plasma, cause coagulation and production of ferment have no action on Hammarsten's fibrinogen, therefore it seems difficult to believe that it is in any way concerned in clotting, since its behaviour is so different from all varieties of plasma. Again, it is well known that the injection of serum—a body containing ferment and paraglobulin in abundance—does not cause intravascular clotting, or, at any rate, only to a minimal extent; and yet with lymph-fibrinogen—a substance which produces no effect on Hammarsten's fibrinogen—we can form a solid clot of the blood in the vessels.

The doctrines of Hammarsten and Schmidt, however true they may be in themselves, can have little or nothing to do with the actual process of coagulation of the blood. For their theory I would substitute the one advocated in this lecture—that coagulation depends on an interaction between proteid-lecithin compounds. Fibrin is produced, not as the result of a special ferment on a special proteid, but as the result of a whole series of substances having many points in common and slight points of difference. In actual coagulation *A-* and *B-fibrinogen* are the two concerned, but we can obtain from the most varied tissues substances greatly resembling these blood-fibrinogens, and which, when added to blood, produce clotting. All these substances contain lecithin, and I think it is impossible to deny that lecithin is essential in the interaction which occurs.

Coagulation is but a type of many other processes, and it is, perhaps, permissible to think the theory advocated in this lecture is better calculated to throw light on these processes than is the fibrinogen, paraglobulin and ferment theory, lecithin and proteid being universally found wherever there is life. Moreover, the fact that plasma by itself, without the intervention of formed elements, is coagulable, tends strongly to put the blood more on the same footing as the protoplasm of a muscular fibre.

I have always spoken above of Hammarsten's fibrinogen, this author having investigated its properties much more completely than Schmidt. Many of the earlier researches on coagulation were carried out by the aid of normal or pathological transudation fluids, under the impression that these represented blood-plasma. So far as normal fluids are concerned, in the horse, dog, and in man this is not the case, the transudation fluid containing either the actual fibrinogen of Hammarsten or a body very closely approaching it, and clotting readily with ferment, not at all with lecithin, and very badly or not at all with leucocytes. This subject, together with many others that I have introduced, I am still investigating; and it must be understood that the chief object of this lecture has been to give a connected account of my own researches.

CLASS V

FURTHER ELABORATION OF OBSERVATIONS
ON COAGULATION

NOTE ON THE RELATION ͏OF THE RED BLOOD-CORPUSCLES TO COAGULATION[1]

It has long been known that injection of 'laky' blood—*i.e.* blood in which the red corpuscles have been dissolved—gives rise to more or less extensive intravascular clotting. This result has been attributed to the action of the dissolved hæmoglobin. It is, however, extremely difficult to understand how this body can have any influence on coagulation, and further information on this point is most desirable. The result of my observations, made at Leipzig, at the Brown Institution, and elsewhere, is to show that the active agent in laky blood is not the hæmoglobin, but the stromata of the red blood-corpuscles—the protoplasmic framework in whose meshes the hæmoglobin is contained.

The mode of experimenting was as follows: Defibrinated dog's blood is mixed with a large quantity of 2 per cent. salt solution, and the corpuscles separated from the serum by rapid rotation in a 'centrifuge.' The corpuscles are again washed with the salt solution by means of the centrifuge, the object, of course, being to get rid of the last traces of serum. The corpuscles are then well shaken up with half per cent. salt solution, and afterwards with a little ether. By this means the red corpuscles are dissolved, and on centrifugalising the fluid so obtained the white corpuscles can be separated.

The supernatant fluid contains hæmoglobin in solution and the stromata, the latter being so much swollen up that they cannot be seen. They are, however, at once made visible and precipitated in a flocculent form by the addition of a weak solution of acid sodium sulphate.

[1] [From the *Practitioner*, March 1, 1886.

By this method [1] the stromata retain their form completely, and can be collected in large quantities, and after washing well with water they can be obtained nearly white.   If the stromata be dissolved in a little alkali and injected into the jugular vein of an animal (I used rabbits), most extensive intravascular clotting is produced; and, provided the quantity be not too small, death occurs almost instantly with general thrombosis of the vessels.   If only a moderate quantity be injected, the blood drawn off after injection clots slowly or not at all.   I have repeated these experiments with stromata many times, and always with a like result.

On the other hand, I have found that the injection of laky blood, freed from the stromata but containing the hæmoglobin, is without any effect on coagulation.   Whether in any circumstances hæmoglobin plays a part in the formation of fibrin I am not prepared to say, but it is rendered extremely doubtful by these observations.

The fact that the stromata cause intravascular clotting is in many ways interesting, and will, I trust, lead to important conclusions regarding the causation of thrombosis during life.   I have recently described [2] a substance which can be obtained from the testis and thymus gland—and, indeed, from many other sources—the injection of a solution of which into the circulation causes extensive intravascular clotting.   This substance is a complex proteid-lecithin compound, and the stromata appear chemically to be largely made up of a similar if not identical body.   I may add that the washed stromata likewise produce coagulation in extra-corporeal blood.

---

[1] For details, see my paper, ‘Zur Chemie der Blutkörperchen,’ in *Du Bois’ Archiv*, 1881.   [Collected Papers, p. 69.]

[2] *Proc. Roy. Soc.* Feb. 4, 1886.   [Collected Papers, p. 135.]

NOTE ON A NEW CONSTITUENT OF
BLOOD SERUM[1]

I WISH in the present note to draw attention to a proteid substance which exists in very small quantity in blood serum. Owing to the difficulty of obtaining a sufficient amount, I shall not attempt to give a complete description of its chemical characters, but shall confine myself chiefly to its physiological properties, which, I venture to suggest, possess considerable interest. It is obtained by rendering undiluted serum distinctly acid by means of dilute acetic or very dilute (4 pro mille) sulphuric acid. Neutralisation does not cause its precipitation; the serum must have a strong acid reaction. It is constantly present in the serum of dog's blood, and when collected by the centrifuge it is precisely similar in physical characters to ordinary fibrin, and only differs from the latter chemically by being more easily soluble in dilute alkali. It is totally different from the soft granular precipitate of paraglobulin, the latter substance being extremely easily soluble in the slightest excess of acid. It is also constantly present in serum of sheep's blood. In the case both of dog's blood and sheep's blood it is only present in very small amount, and in the serum from horse's blood and bullock's blood it was absent in the specimens I have examined. The physiological interest of this substance will be seen from the following.

It is well known that Schmidt regarded two proteid substances as being essential for coagulation. One of these bodies was paraglobulin, a substance existing in large quantity in blood serum. Subsequent investigation has failed to confirm this view, and there can be no doubt that paraglobulin is not

[1] [From the *Proceedings of the Royal Society*, vol. xlii. 1887.]

essential to the process.   But Schmidt has obtained results, the
correctness of which we are in no way entitled to dispute, which
apparently clearly show that the quantity of fibrin formed can
be largely increased by the addition of paraglobulin.   I think
this discrepancy can be explained by the help of this new
substance, and this will be best shown by describing the following experiments :—

Two portions of peptone-plasma were taken, and
To No. 1, an equal quantity of sheep's serum was added ;
To No. 2, a small quantity of a solution of the new substance.
No. 1, after many hours only presented a scarcely perceptible
flocculus of fibrin.

No. 2 was quite solid in fifteen minutes; on squeezing out
the fluid from the clot, and again adding a solution of the new
substance, the mixture again clotted through and through.

Now Schmidt's experiments were very much of this nature.
He found in certain specimens of hydrocele fluid that the
addition of fibrin ferment produced very slight clotting, whereas
on the further addition of a substance which he regarded as
paraglobulin a decided clotting took place.   Now sheep's serum
contains plenty of paraglobulin and plenty of fibrin ferment,
but it has no appreciable effect in my experiments.

But this new substance, which, it must be remembered, is
only present in very small quantity in serum, had the most
marked influence, and hence I conclude that it is the new
substance, and not paraglobulin, which increases the amount of
fibrin.   It may be mentioned that in preparing paraglobulin a
certain amount of the new substance is always precipitated with
the former substance.

A second physiological property of this new substance is the
effect it exerts when injected into the circulation of a living
animal.

It is very exceptional to find that the injection of blood
serum produces any effect, serum containing plenty of paraglobulin and ferment, but only traces of the new substance.

But the injection of a solution of this body prevents the

coagulation of the shed blood. Occasionally, as the result of the injection, very small thrombi are formed; possibly, if more could be obtained, considerable intravascular clotting might be set up.

The following is an example :—

A quantity of the new substance obtained from 300 c.c. sheep's serum and well washed was dissolved in dilute alkali and salt solution. (The amount of substance was, I estimate, 0·2 grm.) This solution was injected into the jugular vein of a rabbit. The blood of this rabbit previous to the injection clotted in two minutes; after the injection the blood drawn off remained quite fluid for three hours—time of observation. It clotted, however, directly on adding some of the solution injected.

The injection of considerable quantities of serum or of paraglobulin I have not found to have any appreciable effect.

Of itself this substance, since it exists in so small amount, is of little interest, but as it appears to vary in quantity in different animals and under different circumstances, it is easy to see that misapprehensions as to the influence of paraglobulin on coagulation might easily arise.

These observations also throw great doubt on the power of fibrin ferment to produce a so-called intoxication.

This substance has an extremely feeble influence on dilute $MgSO_4$ plasma, and hence contains but a trace of fibrin ferment. Since it is closely related to the fibrin-yielding matters of the plasma, and to the tissue-fibrinogens I have elsewhere described, I should propose to call it ' serum-fibrinogen.'

BLOOD-PLASMA AS PROTOPLASMA

An extract from the Arris and Gale Lectures, delivered at the Royal
College of Surgeons, June 1886.

THERE can be no question that there exists a very great analogy
between the coagulation of the blood and rigor mortis. Rigor
mortis can hardly be looked upon otherwise than as the last
vital manifestation of muscle, and hence there are sound reasons
for looking on coagulation as the last act of the living blood.
No one doubts for an instant that muscle-plasma is living protoplasm, and the only very strong reason which would prevent
us from adopting blood-plasma as protoplasma is the supposition
which exists relative to the participation of recognised form-
elements of the blood in the process of coagulation.

Admitting that coagulation is a 'vital' act, so long as it is
apparently necessary that form-elements should intervene to
effect this process, so long are we justified in doubting the right
of the plasma to the epithet 'living.' But the idea that form-
elements do participate in coagulation, and indeed are necessary factors in the process, has taken very deep root amidst
physiological doctrines, and if we are successfully to uphold the
claims of the plasma to be fluid protoplasm, we must first show
that this doctrine is unfounded.

A very prevalent belief is that the white blood-corpuscles
are important agents in the process. The acceptance of this
view is chiefly due to the experiments of Alexander Schmidt,
and in the main his experiments are as follows. By allowing
blood from the vein of a horse to flow into long vessels surrounded by ice, the coagulation of the blood is greatly delayed
and the corpuscles are able to sink. The red corpuscles sink
more readily than the white, the latter partly forming a layer on
the surface of the column of red, and partly being scattered

through the lower layers of the plasma. Ultimately, in spite of the cold, coagulation takes place, and it is then observed that clotting begins earlier and is most complete in those strata of the plasma where there are most white corpuscles. Again, by filtering the plasma at a low temperature, before coagulation has had time to occur, the complete removal of the white corpuscles can be effected, and the plasma so obtained is found to clot very slowly and scantily, even at ordinary temperatures.

Relying chiefly on these experiments, Schmidt was led to the conclusion that white blood-corpuscles were essentially concerned in coagulation. But, as I shall now point out, this conclusion is totally unjustified. In another form of plasma, and more suited to experiments of this nature, to which I shall afterwards frequently have to allude, I have found that simply cooling down the plasma suffices to give rise to a marked precipitate. A substance separates from the plasma on cooling, and it can be proved by the most exact experimentation that this very substance is of the greatest importance in coagulation; in fact, it plays the part attributed by Schmidt to the white corpuscles. And in Schmidt's experiments there can be no doubt, since all the operations were carried out at a low temperature, that in removing the white corpuscles this new substance was at the same time removed. Indeed, Schmidt himself describes the existence, side by side with white corpuscles, of a large quantity of granular matter. In the one case this sinks with the white corpuscles, in the other it is removed by filtration. Schmidt supposes that the cold prevents the white blood-corpuscles from breaking up, but not entirely so, and he looks upon this granular matter as being *débris* of white corpuscles which have gone to pieces in spite of the cold. If this were so his conclusions might be justifiable, but it is a pure assumption on his part; and in face of the fact that in other forms of plasma, perfectly free from any corpuscles, a granular precipitate arises on cooling, it is obvious that Schmidt's conclusions as to the participation of the white corpuscles were not justified.

Since the publication of these fundamental researches of Schmidt, in which he endeavoured to show that the absence of white corpuscles renders the coagulation very slow and very imperfect, an apparent confirmation of his views has been afforded by experiments of a more direct nature.

I first showed that, by adding to plasma free from cells isolated lymph-corpuscles obtained from lymphatic glands, coagulation at once ensues. I myself at this time looked upon this as a confirmation of Schmidt's views. But it is obviously only a proof if lymph-corpuscles are identical with white blood-corpuscles; and this I believe not to be the case. No doubt the white blood-corpuscles are to a great extent recruited from the lymph-corpuscles; but I have described experiments which show that when once a lymph-corpuscle has got into the circulating blood-plasma, it loses all its power of inducing coagulation. It is, then, extremely doubtful whether white blood-corpuscles have anything to do with coagulation; and, on the other hand, it is certain that plasma free from all corpuscular elements will clot spontaneously. This latter statement must now be considered in greater detail.

The Coagulation of Plasma without the Participation of any Cellular Elements

This can be most advantageously studied in a form of plasma to which I shall allude in the future as 'peptone-plasma.' It is obtained by the injection of a solution of peptone into the circulation of a dog. A few minutes later the animal is bled to death; the blood does not clot, and by means of the centrifugal machine plasma can be obtained perfectly free from all form-elements. This plasma is spontaneously coagulable, or, more exactly, it will clot, yielding normal fibrin without the addition of anything which can be in any way regarded as a fibrin factor. The injection of peptone prevents the normal coagulation from taking place. The plasma is, however, coagulable. The process only requires to be started by some simple chemical or mechanical stimulus; just as muscle is contractile, but requires to be set going by a stimulus before it actually contracts. Among the means by which the plasma can be made to coagulate are neutralisation with acetic acid or carbonic acid, dilution with water, or even salt solution. Mechanical means, as filtering through a clay cell, or even in some cases through filter-paper, are also effectual.

Now filtration through a clay cell is not a ' fibrin factor'; acetic acid is not a ' fibrin factor'; they are stimulants applied to the blood-plasma, which answers by clotting. Peptone-plasma is not the only instance in which mechanical stimulation produces clotting, for I think it must be admitted that the plasma in the vessels clots on mechanical stimulation. It is well known that if a thread be drawn through a vessel it will become covered with clot. This cannot be attributed to the presence of a foreign body, for Zahn has shown conclusively that a foreign body only produces coagulation when it incompletely blocks the lumen of a vessel. The blood-current must continue. Around a globule of mercury, completely blocking the lumen of a vessel, clotting does not occur ; but if it only partly obstructs the blood-current, it gives rise to a thrombus. Zahn attributes the clotting which occurs under these circumstances to the white corpuscles. The latter, he thinks, attach themselves to the foreign body, and by their disintegration yield fibrin. If the blood-current is completely stopped, enough white blood-cor-puscles will not present themselves, so that no thrombus will be formed ; but if a certain amount of flow can go on, enough corpuscles will be brought to the spot and a thrombus formed.

But in view of the fact that mechanical stimulation—*e.g.* sucking through narrow pores—suffices to induce the coagulation of peptone-plasma entirely free from form-elements, Zahn's explanation is open to considerable doubt. Moreover, the micro-scopical observations on which he relied admit of another explanation. Rindfleisch also remarks, as the result of his anatomical observations, that thrombi are especially liable to be formed, under otherwise favourable circumstances, at those parts of the vascular system which from their configuration are prone to give rise to eddies and sudden changes in the flow of blood, and that pathological changes in the vascular wall are not con-stantly associated with thrombosis, provided no projecting pro-cesses or marked unevenness of the diseased surface be present.

To return to peptone-plasma. I have stated and explained
its spontaneous coagulability. This power is, however, entirely
lost if we cool down the plasma to a temperature of about 0° and
maintain it at this temperature for some time. As the result of
the cooling a precipitate is formed, and with the removal of this
precipitate the power of spontaneous coagulation is lost. The
plasma will still clot, will still yield large quantities of fibrin,
but only on the addition of bodies which must be looked upon
as fibrin factors. I must call special attention to this precipitate.
The most important point in connection with it is its peculiar
microscopic appearance. Chemically it is a very complex
proteid substance, but when examined microscopically it appears,
not in the form of irregular granular masses, as is generally the
case with proteid precipitates, but in the form of perfectly
regular round discs. These discs are three or four times smaller
than red corpuscles. Their discoidal character is readily seen
when they roll over. After they have been separated for some
time they run together into masses and finally form granular
heaps. When first separated they will redissolve on warming,
and if this process be observed under the microscope interesting
observations may be made. In a preparation examined immediately after the removal of plasma from the ice we see the
discs, some isolated, others collected in groups of four or five
hanging together, so as to form rosettes. If one of these
rosettes be kept in view for a few minutes, whilst the temperature of the plasma becomes raised to that of the warm room, it
will be seen that its outlines are changing, that an irregular
angular lump is forming, and that this lump is gradually
becoming round. Finally a round disc is formed, differing only
from the other discs in being larger. It may be as large as a
red blood-corpuscle. These discs are distinctly biconcave, the
larger ones being, so far as form is concerned, quite indistinguishable from red blood-corpuscles. In the smaller ones it
is not quite so easy to make out the central depression, high
powers being necessary.

These changes can only be observed in the case of a plasma
thoroughly under the influence of peptone and shortly after the

separation of the substance. In other cases the discs rapidly run together into irregular granular masses.

There is no doubt that this body separable by cold is identical with the so-called ' Blutplättchen ' of Bizzozero. The latter are, according to all observers, plentifully present in peptone-blood, and these unquestioned Blutplättchen are absolutely indistinguishable in every respect from the discs which separate from the plasma on cooling.

Blutplättchen are regarded by all observers who have attacked the problem as definite form-elements, by which I understand them to mean organised bodies ; and it is difficult to know how else to regard them, for they have just as much right to be regarded as a form-element as a red corpuscle has. Many attribute to these bodies, in regard to coagulation, the powers which Schmidt has referred to the white corpuscles. I think they are perfectly right, but they leave the question practically where it was before. Whether this or that form-element must necessarily step in to help on or to initiate the powers of coagulation is, for the problem we are concerned with, practically without importance. It is obvious that if the blood-plasma cannot clot without the intervention of form-elements, be they white corpuscles or be they the structures called ' Blutplättchen,' our contention that the plasma is ' living ' has but little to support it. But we have seen that form-elements are not necessary ; that Blutplättchen have a fluid representative in the plasma. Blutplättchen are only form-elements at a certain temperature.

The substance separable by cold, the microscopical characters of which have just been described, is dissolved in the alkali of the plasma. The lower the temperature the less quantity of the substance can remain in solution. A very slight cooling suffices to precipitate a certain quantity. The substance can be also precipitated by acids, and by certain strengths of neutral salts.[1]

---

[1] Hence there are two kinds of ' salt-plasma ' : the one obtained by using strong sulphate of magnesia solution, containing none of this substance, owing to its having been precipitated ; the other by using more dilute salt solutions, still containing this substance and clotting on dilution.

The Fibrin Ferment

It has also been pointed out that peptone-plasma is spontaneously coagulable only so long as the peculiar body we have just described is present. But this coagulation presents a very special and important feature, for when. it occurs not only is fibrin formed, but at the same time a large amount of fibrin ferment makes its appearance.

The fibrin ferment is a substance capable of converting solutions of fibrinogen into fibrin. We know it is not a proteid, and we know nothing more about its chemical nature except that it is destroyed by boiling.

There is no ferment in peptone-plasma, but if the plasma be made to clot by the means described above (CO_2, &c. &c.), the serum which exudes from the clot is found to contain large quantities of the ferment.

Both clotting and fibrin formation are dependent on the presence of the body separable by cold. The more of this substance there is present, the more easily does clotting occur and the more ferment is found. With the gradual removal of this substance clotting becomes less and less easy, and less and less ferment is formed. I attribute very great importance to this coincident formation of fibrin ferment. It would be absurd to contend that every process of fibrin formation is of a vital character. It is on a consideration of the whole of the phenomena, as seen in the plasma, that I base my claims of vitality for the blood-plasma.

It would appear extremely probable, from what we know of the subject, that ferments make their appearance as the result of protoplasmic disintegration or death. Take, for instance, the pancreas. We here get, in the interval of rest, a storing up of definite granules, which, under the influence of activity, are discharged or disappear in the form of secretion. There is further reason to believe that these granules are eminently of a protoplasmic nature, eminently living (Ogata). Previous to the appearance of ferment these granules must undergo fundamental change, for in a fresh gland, no matter how much 'zymogen,' as it is called, be present, there is no free ferment. But in a gland in which post-mortem changes have taken place ferment is present, and this change can be accelerated by artificial means.

How very marked is the analogy between the case of the pancreas cell and the blood-plasma! Trypsin is formed as the result of the disintegration of this cell, fibrin ferment as the result of that of the blood-plasma. The capability of forming a large amount of trypsin goes hand-in-hand with the presence of a large amount of granules in the inner zone of the pancreas cell. The capability of forming a large amount of fibrin ferment depends on the amount of a substance which, as soon as it appears at all, appears in the form of the most definite ' granules ' conceivable. I may here call attention to the fact that fibrin ferment appears to be the post-mortem ferment of muscle—a fact which greatly strengthens the analogy between rigor mortis and coagulation of the blood.

Histologists have distinguished in the protoplasm of cells two distinct constituents : the one, formed, generally appearing in granules, and designated ' protoplasma ' in a restricted sense ; the other, more fluid, unformed, and called ' paraplasma' (Kupfer). The formed constituents appear to be more especially concerned in the functional activity of the cell, hence the denomination ' protoplasma ' (in a special sense). As examples of cases in which ' granules ' are obviously the functionally active part of a cell, we might instance the chlorophyll granules of the vegetable cell and the granules which are so obvious in many secreting cells. The nearly universal occurrence of such granules in cells, and the strong reasons there are for regarding them as the most actively living part of the cell, have been shown in a recent important work of Altmann.[1]

Now, in the case of the blood-plasma, we might regard the substance which separates from the plasma on cooling as the more special protoplasma, the rest of the plasma being the paraplasma. The power of spontaneous coagulation, the power of giving rise to fibrin ferment, depends on a special substance, which, when it is made visible, appears in the form of the most definite ' granules.'

[1] *Ueber die Zelle.*

The Vascular Wall and the Blood

Genetically considered, both blood and endothelium are differentiations of one and the same protoplasmic mass. The blood is merely the more fluid central part of the originally solid protoplasmic cord. The blood and the vascular wall may then be looked upon as a protoplasmic unit. That the vascular wall exerts a great influence on the blood is evident from the fact that the blood undergoes changes which finally terminate in coagulation, so soon as it leaves the vascular wall. And, similarly, we know that a motor nerve separated from its ganglion cell undergoes degenerative changes and death.

The change in the blood takes place much more quickly than the change in the nerve, and its nature will become more evident from what follows. The great point on which I wish to insist is the fact that it is the plasma which undergoes change. Previous authors have always supposed that the change took place in the form-elements. Thus Schmidt supposed there was a 'death' and breaking up of the white corpuscles. But the fundamental change is a change of the blood-plasma itself. The blood-plasma may be looked upon as the most sensitive part of the blood.

The exact nature of this change I do not know, but that it does occur may be proved in the following way. Peptone-blood clots with leucocytes from lymphatic glands, and it is the plasma of the blood which has this power of clotting with leucocytes. But blood which has not left the vessels, be it ordinary normal blood or peptone-blood, does not clot with these leucocytes, hence it is obvious that the plasma must undergo some change when it leaves the vascular wall. A small quantity of leucocytes added to peptone-blood immediately after its removal from the vessel will cause rapid clotting; but these same leucocytes do not cause clotting when injected in large quantity either into the general circulation or into an isolated vein. It is the plasma, then, which so rapidly undergoes change when the blood leaves the vascular wall. This great sensitiveness is,

I contend, another argument in favour of the protoplasmic nature of the plasma.

Before leaving this division of our subject, it may be well to allude to the relation of the red corpuscles to the plasma. We have seen that the blood-vessel and the plasma form what I may call a ' protoplasmic unit,' and we have evidence to show that the red corpuscles are also integral parts of this endothelium and plasma union. I must admit, however, that this point is still extremely obscure. It appears, however, pretty certain that the red corpuscles of shed blood are not quite the same things as the circulating red corpuscles. It has been shown by transfusion experiments that the corpuscles of defibrinated blood are incapable of persisting in the organism, that they very speedily disappear, and that defibrinated blood has no more lasting effect in making up for a loss of blood than salt solution has. It would, therefore, appear that in the death of the blood the red corpuscle also suffers, though in a less marked and less easily demonstrable manner than does the plasma.

Some Chemical Characters of the Red Corpuscles and Plasma

A consideration of the chemical structure of the red corpuscles is of interest in regard to the question before us. It is common to speak of the red blood-corpuscles as consisting of hæmoglobin, albuminous stroma, and salts. But this is only the case in the same way as chlorine and sodium are constituents of common salt. It is impossible to avoid the conclusion that the red blood-corpuscle is a chemical unit, that we must speak of a red blood-corpuscle substance, although capable no doubt of being split up into smaller molecules such as hæmoglobin and stroma, as being as decided a chemical individual as sodium chloride, and for the following reasons. Hæmoglobin cannot be present as such, for it is extremely soluble in plasma, and, if free in the blood, becomes converted into methæmoglobin and excreted by the kidney. Under ordinary circumstances, it need hardly be said that the hæmoglobin does not pass into the plasma; and this cannot be explained by supposing that the stroma forms a protective envelope for the hæmoglobin, since the stroma itself, if free in the blood-plasma, is a most dangerous poison, producing wide-spread intravascular clotting. I think these considerations drive us to the conclusion that the hæmoglobin and the stroma are in chemical union to form the 'red blood-corpuscle substance.' All the evidence we possess tends to the conclusion that the red blood-corpuscles are protoplasmic structures, and I think the chemical conditions I have described well illustrate the chemical relations of living matter.

Pflüger first suggested the idea of a 'protoplasmic molecule,' and the evidence, so far as the red corpuscle is concerned, certainly points to the correctness of this idea. One can obtain from red corpuscles (and from other tissues) various extremely complex bodies, and these are cited as being the chemical constituents of the tissue. But they do not exist as such side by side; there is a chemical union.

A similar state of things prevails with regard to the plasma. Thus, a body (fibrinogen) can be isolated from the plasma, which will clot readily with fibrin ferment, yielding fibrin, and will coagulate by heat on warming to 52·55° C. But plasma itself (peptone-plasma and plasma in the vessels) will not clot with fibrin ferment, neither will plasma coagulate, proper pre-cautions being taken, till very high temperatures are reached (over 90° C.); plasma behaving quite differently in this respect to serum—a dead fluid. Plasma which has been artificially altered and changed will clot with fibrin ferment, but the plasma in the vessels, and plasma altered as little as possible after leaving the vessels, are quite unaffected by fibrin ferment. For this and other reasons which I cannot adduce here, there can be no doubt that though a definite substance, fibrinogen,[1] can be obtained from plasma, it is not present as such; it is a decomposition product of plasma substance, and plasma substance is not coagulable with fibrin ferment.

Now let us consider the close analogy between this case and the pancreas. The pancreatic ferment trypsin arises as the result of the death of the pancreatic protoplasm, and just as fibrin

[1] Fibrinogen of Hammarsten.

ferment is incapable of attacking living blood-plasma, so pancreatic ferment is incapable of attacking living pancreatic protoplasm and living intestinal protoplasm.   Fibrin ferment acts on a special body—fibrinogen.   The reason why it does not act on plasma is that there is no fibrinogen there.   Trypsin acts on proteids; it attacks a proteid molecule; it cannot touch a protoplasma molecule any more than fibrin ferment can touch a plasma molecule.

The Chemical Processes in Coagulation and the Chemical Processes of Life

I shall treat this subject very briefly, as I have discussed it at much greater detail elsewhere.[1]   It is well known that the existing doctrines on coagulation explain it as an essentially fermentative process.   From my own researches I am entirely opposed to this explanation.   I have alluded to the fact that we may distinguish in the blood-plasma a part we may call the ' special protoplasma,' and another part the ' paraplasma.'   The essential process in coagulation is an interaction between protoplasma and paraplasma.   Both protoplasma and paraplasma consist of proteid and lecithin; that is to say, these bodies can be obtained from them.   In the process of coagulation there is a lecithin transference, the protoplasmic moiety losing, the paraplasmic moiety gaining lecithin, the result being the formation of fibrin, also a proteid-lecithin mixture or compound.

I only wish to say that this is the main feature in the process; the actual chemical process is, no doubt, one of enormous complexity.   It is the break up of the plasma molecule and the new formation of endless other smaller molecules (fibrin constituents of serum).   It is the passage from life to death.   We cannot hope to explain all the chemical processes which accompany that tremendous change, but it is definitely made out that, so far as the formation of fibrin is concerned, the process is a lecithin-proteid interchange.

Coagulation of the blood is probably but the type of a

[1] Croonian Lecture, Royal Society, April 8, 1886. [Collected Papers, p. 141.]

process occurring everywhere in protoplasm; or, in other words, many of the phenomena of life are fundamentally processes of the nature of coagulation. Previous investigators have regarded coagulation as essentially a fermentative process, the action of a special ferment, fibrin ferment, on a special proteid, fibrinogen; and the prevailing view as to the nature of the chemical processes of life is that they are in the main fermentative processes. But what we know of the blood, perhaps the one tissue which we can thoroughly get at experimentally, is not favourable to this—the existing—view.

The blood ferment is essentially a post-mortem production, the result of the final explosion of the plasma molecule. No doubt fibrin ferment is capable of causing the appearance of fibrin. It will convert a dead and isolated constituent of the plasma into fibrin, but it will not make plasma itself clot. (I refer to plasma in the vessels and peptone-plasma.)

The injection of a large quantity of fibrin ferment leaves the animal unharmed. It can then exert but little influence on the life of the blood, but, on the other hand, we can exercise a tremendous influence on the blood by the injection of a certain proteid-lecithin compound.[1] This is not a ferment; it does not act in the least like a ferment, but it will clot the blood in the vessels from one end of the body to the other, provided enough be injected. The amount of intravascular coagulation is entirely dependent on the quantity of substance injected. The substance is used up in the process and enters into the formation of the fibrin produced. The lecithin it contains is essential to its action.

If, then, the process of coagulation is rightly regarded as typical of many processes of life, these processes are not catalytic processes. They are the result of the union in definite proportions of complex proteid-lecithin compounds, it being remembered that lecithin and proteid are constant wherever there is life.

As in the case of the blood, ferments may arise as by-

---

[1] This substance is very widely distributed. It can be obtained from lymph-glands, chyle, brain, testis, thymus, &c.

products, but they are effects, not causes, of the main processes. That lecithin is profoundly concerned in the formation of these ferments is certain so far as the blood is concerned.

Conclusion

In the preceding pages I have endeavoured to show what great claims the blood-plasma has to be regarded as living matter. One is accustomed to look upon life as always associated with form. The plasma is not formed, it is formable.

Let me once more draw attention to the peculiar precipitate produced in the plasma by cold. When once this precipitate is visible, it is very difficult to look on it as other than a form-element. Indeed, it has been seen and described as a form-element, and it has just the same right to this title as a red corpuscle has. The morphological characters of the precipitate are identical with those of the red corpuscles, and to this must be added the interesting fact that, chemically, the stroma of the red blood-corpuscles—that is to say, that part of the red corpuscles which has not become differentiated into hæmoglobin—is identical with the precipitate produced by cold. Moreover, this latter contains iron in an organic form. Now the development of the red corpuscles is profoundly obscure; overwhelming differences exist between the various authors who have investigated the question, and I put it, in face of what I have found and described above, as a suggestion, Are red blood-corpuscles deposits from the plasma?

The substance separable by cold may be regarded from two different aspects. On the one hand, the process is evidently allied to crystallisation, and the discs might be regarded as a crystalline precipitate; on the other hand, they may be looked upon as imperfect cells. *Omnis cellula e cellula* is a dogma. Is it true? Are blood-corpuscles and plasma-discs cells, or are they crystals? Surely, if ever we shall have a borderland between 'vital' processes and ordinary physical and chemical processes, we shall find it in the blood-plasma.

A THEORY OF THE COAGULATION OF THE BLOOD [1]

IT is now several years since I commenced to investigate the coagulation of the blood in the laboratory of my revered teacher, Professor Ludwig. I have since then devoted my attention to the question, and from time to time published short notices of the results of my work. I wish now to give a collected account of my scattered communications, although this cannot be in any way exhaustive, since I am still far from a complete understanding of the complicated phenomena.

It is possible, however, to show that the previous theories of the subject are inadequate, and that the explanation must be sought in quite another direction.

It is generally assumed that in the coagulation of the blood the active help of the formed elements of the blood is necessary.

It is said that at the moment of coagulation certain substances which are wanting in the plasma are supplied to it by the formed elements. Some authorities look upon the white blood-corpuscles as the active factors, others a special kind of corpuscles which have been designated ' *Blutplättchen* ' (blood-platelets).

I shall first attempt to prove that this theory is only true, if at all, in a very limited sense, and that *the blood-plasma itself, free from all formed elements, contains all that is necessary for coagulation.*

It will be advantageous to consider the behaviour of four sorts of plasma which are used in experiments on the blood—peptone-plasma, weak salt-plasma, strong salt-plasma, and cooled plasma.

Peptone-plasma, as is well known, is got by bleeding an animal into whose veins a solution of peptone has been injected,

[1] [Translated from ' Beiträge zur Physiologie,' Festschrift für Carl Ludwig. Leipzig, 1887, p. 221.]

by which means the blood is prevented from coagulating. This blood is then centrifugalised.   In this way the plasma is separated from the corpuscles, and can be siphoned off as a clear fluid free from any trace of formed elements.

In spite of this absence of formed elements, the plasma may be said to be spontaneously coagulable, since it coagulates, as Fano has shown, when submitted to the action of a stream of CO_2, or when diluted with water or filtered through a clay cell.

No one can regard these influences as 'fibrin-factors'; rather should they be regarded as increasing the tendency of the plasma constituents towards dissolution.   The natural process of clotting is hindered by the injection of peptone, and these influences, acting in a reverse direction, neutralise and overcome this hindrance.

If, however, the peptone-plasma is cooled for some time to 0° C. it can no longer be made to clot by these simple means. Under these conditions a precipitate is formed in the plasma, and, with the precipitate, the plasma loses its power of spontaneous coagulation—i.e. of forming fibrin at the expense of its own substance.

We may here pause a moment to consider Alexander Schmidt's classical experiment on cooled plasma.   Schmidt received blood from the vein of a horse into vessels surrounded by ice.   In this way clotting was prevented, and the corpuscles settled gradually to the bottom of the liquid.   The plasma was siphoned off and filtered through filter-paper, being kept all the while carefully at 0° C.   The clear plasma obtained by this means shows little or no tendency to coagulate, and Schmidt ascribes this to the absence of the white blood-corpuscles.

It is, indeed, very probable that in this experiment the white blood-corpuscles are removed ; but we know, from our experience with peptone-plasma, that the consequence of cooling the plasma to 0° is a precipitate which is of the highest possible importance for the spontaneous coagulability of the plasma.   This precipitate would remain with the leucocytes on the filter, so we see that Schmidt's results by no means prove that leucocytes are necessary to coagulation.

For a second experiment Schmidt simply allowed the cooled blood to stand. The red corpuscles sink to the bottom, and over them is a layer of white corpuscles. Clotting takes place before all the white corpuscles have sunk, and we find that it begins sooner, and is most complete, in the deeper strata of plasma, where the white corpuscles are most numerous.

But Schmidt himself mentions the fact that a large quantity of granular matter is present with the white corpuscles. He regards the granules as the *débris* of broken-down corpuscles, without, however, adducing any special argument for his belief. It seems most probable that these granules are identical with the precipitate which we have seen to occur in plasma in consequence of cooling.

Weak salt-plasma is got by centrifugalising blood which has been kept from clotting by receiving it direct from the vein of the animal into an equal volume of 10 per cent. solution of NaCl. The plasma thus contains 5 per cent. of NaCl. It clots on dilution with water.

Strong salt-plasma is obtained from blood which has been kept from clotting by receiving it into a saturated solution of $MgSO_4$ in the proportion of three or four parts of blood to one of $MgSO_4$ solution. This plasma does *not* clot on dilution, but requires the addition to it of fibrin ferment.

To explain this difference in the coagulabilities of the two kinds of plasma two assumptions have been made, neither of which are justified by facts : these are, that the strong $MgSO_4$ solution has no effect on the plasma, but that it prevents the escape of the fibrin ferment from the cells, while the weak NaCl solution is powerless to do so.

If to peptone-plasma, which clots on dilution, we add a saturated solution of $MgSO_4$ in the proportion of one part of the solution to four of the plasma, we find, either immediately or after a short time, that a granular precipitate is formed in the fluid. With the removal of this precipitate the plasma loses its power of spontaneous coagulation. It clots no longer on dilution, but does so on the addition of fibrin ferment, just as ordinary $MgSO_4$ plasma does.

Hence it follows that the addition of $MgSO_4$ is not without effect on the plasma, but, just as cooling does, precipitates some substance, on the presence of which depends the spontaneous coagulability of the plasma. This precipitate is in all respects similar to that produced by the action of cold. The belief so generally held that plasma by itself is unable to clot, rests mainly on the above-quoted experiments of Schmidt and on the fact that $MgSO_4$ plasma is not spontaneously coagulable.

The fact was overlooked that both methods of obtaining the plasma were necessarily attended with the loss of a constituent of the plasma which was of the highest importance for its coagulation.

Another proof has been brought forward of the participation of the white blood-corpuscles in the act of coagulation, and this is the fact that leucocytes obtained from a lymph-gland added to a plasma cause clotting. But it can be shown that the lymph-corpuscles on entering the circulation undergo certain changes, so that the action of the leucocyte circulating in the blood-plasma cannot be regarded as similar to that of the leucocyte obtained from a lymph-gland. The following experiments show this:—

A dog was injected with a solution of peptone. Some blood that was drawn off did not clot spontaneously, but clotted quickly on the addition of lymph-corpuscles. A large quantity of the same corpuscles was now injected into the vein of the dog without producing the slightest effect. Shortly afterwards some blood was again drawn off. This specimen also would not coagulate until some lymph-corpuscles were added to it, when it coagulated immediately. This experiment was many times repeated, but always with the same result. In no case was there a trace of intravascular clotting.

In a second series of experiments, the lymph-corpuscles were injected into the dog without any preliminary injection of peptone. The only result was that blood drawn off after the injection coagulated more slowly. For the most part, however, this retardation of the clotting was but slight. In this case, also, addition of lymph-corpuscles caused immediate clotting.

The injected lymph-corpuscles have no action on the peptonised dogs, and take no part in intravascular clotting. The blood, too, which is drawn off after the injection, and which must contain these corpuscles, does not clot, so that it is evident that the lymph-corpuscles experience some change on their entry into circulating blood, whereby their power of causing coagulation is lost. Thus we are justified in coming to the conclusion that no satisfactory proof has yet been afforded of the necessary participation of the leucocytes in coagulation. The question whether any other formed elements take part in coagulation will be better discussed in the next section.

THE PROCESS OF COAGULATION

My experiments have shown that the clotting of the plasma is the result of the interaction of two bodies. These bodies are either compounds or mixtures of proteid substances and lecithin, and the latter substance seems to be the essential factor. It is proposed to call these two precursors of fibrin ' A- and B-fibrinogen.' In determining the properties of these two bodies there are great difficulties to contend with. It is easy enough, indeed, to separate them in various ways from the plasma. In the process of separation, however, they undergo slight changes, so that it becomes necessary to deduce their properties from the behaviour of a plasma in which both substances are present.

A-fibrinogen

This is the body which is precipitated from peptone-plasma under the influence of cold. It forms only a small fraction of the coagulable constituents of the plasma. Slight cooling of the plasma suffices to precipitate some of this, but for its complete separation long standing in ice is necessary. When freshly precipitated it is soluble in 4 per cent. salt solution and dilute alkalies, as well as in warm plasma. After standing some time it is no longer soluble in warmed plasma, but only swells up, and if separated by means of the centrifuge, it appears as a thin layer, much like fibrin. This layer is hardly or not at all soluble in 4 per cent. salt solution, and is very sparingly soluble in weak alkalies. It is to outward appearances nearly identical with fibrin, being, however, of a more mucous consistency. After pressing it between filter-paper, it forms a little lump which is not to be distinguished from fibrin, save that it is almost entirely dissolved in a 0·2 per cent. solution of HCl. The microscopical appearance of this substance is very interesting. Soon after formation the precipitate consists of small round plates which hang together in groups. On warming they melt together to form larger plates, which in appearance are very similar to red blood-corpuscles, though, of course, they are colourless. After standing for some time these plates break down into granular masses.

It is a question whether this body is to be regarded as a precipitate from the plasma or as a formed element of the blood. If we assume the latter, we must ascribe to these formed elements the peculiarity of only appearing in the cold, while on warming they are either dissolved or swell up so as to become invisible. It might also be thought that part of the substance was already present in the warm plasma, but was precipitated together with the corpuscles by the centrifuge. Of the formed elements of the blood that have been described, it is most like the ' *Blut-plättchen*.' Yet it is difficult to come to a decision, as under this name many different kinds of substances are understood, so that it is not possible to say whether all observers had to do with the same structures.

A-fibrinogen may be precipitated not only by means of cold, but also by addition of $MgSO_4$. Exactly so much of a saturated solution is added to the peptone-plasma as not to produce any *immediate* precipitation. Generally, we can add one volume $MgSO_4$ to four volumes of plasma. After standing for a short time at ordinary temperatures a precipitate is produced which at first is resoluble in dilute salt solution, but loses its solubility after it has been separated for some time. It reacts in the same way with dilute alkalies. It shows all the properties of the precipitate which is produced by cold. The A-fibrinogen is the substance which in consequence of its great instability imparts to the plasma the property of ' spontaneous coagulability.'   Peptone-plasma, or 4 per cent. salt solution plasma, is coagulated by the means already mentioned (influence of CO_2, dilution with H_2O, filtration through a clay cell) only when it contains this substance.

If, however, the A-fibrinogen be removed by cooling or addition of $MgSO_4$, the addition of a real fibrin-factor is required for the production of fibrin, as we shall describe later. Fibrin ferment has *no* effect on A-fibrinogen.   If fresh peptone-plasma is treated with ferment no clotting takes place, and on cooling, the precipitate of A-fibrinogen is thrown down in the usual way.   On the other hand, fresh peptone-plasma clots with lymph-corpuscles.   In whatever manner, however, peptone or salt-plasma be made to clot, the A-fibrinogen disappears each time in the process, so that there can be no doubt of the participation of this substance in the process.

B-fibrinogen

If the A-fibrinogen is separated by cooling or by addition of $MgSO_4$, the plasma must still contain in all cases plenty of substances capable of giving rise to fibrin, for it is possible, by proceeding according to Hammarsten's method, to get from the plasma a large quantity of a substance which is in all points identical with the classical fibrinogen of Schmidt and Hammarsten.   But the plasma itself shows properties which differ entirely from a solution of the fibrinogen of Hammarsten.   The plasma clots very easily with lymph-corpuscles, and after this clotting no more fibrinogen can be extracted from the plasma. The plasma clots neither with fibrin ferment nor with serum. Hammarsten's fibrinogen, on the contrary, clots with the utmost ease with ferment or serum, and will not clot on the addition of leucocytes obtained from lymph-glands.   Thus we see that Hammarsten's fibrinogen cannot exist as such in the plasma, but must be regarded more as a product of some chemical change. This view is supported by the following experiments:—

If to plasma, which is free from fibrinogen, we add sodium chloride to about half its point of saturation, a finely-flocculent precipitate (1) is produced, which, when collected and pressed between filter-paper, is easily soluble in dilute salt solution. A solution of this precipitate (1) clots either with ferment or with leucocytes. If the solution is precipitated again by addition of salt, we get a precipitate (2) which is much coarser and more like fibrin. It is also much more slowly dissolved by dilute salt solution than precipitate (1). The solution of precipitate (2) clots only with ferment, and not at all with leucocytes.

Thus we see that this fibrinogen of Hammarsten has a fore-runner or ' mother substance ' in plasma, which possesses different properties, and this substance I call ' B-fibrinogen.' Experiment shows, further, that the change takes place by a transition stage [solution of precipitate (1)], which shows properties mid-way between the mother substance and Hammarsten's fibrinogen. How exceedingly unstable the constitution of this body is, is shown by the fact that we have only to dilute the precipitate (1), or even the plasma, sufficiently, in order to bring about the change into the ordinary fibrinogen. Thus magnesium sulphate plasma, after dilution with ten times its bulk of water, will clot only with ferment and not with lymph-cells.

Instead of the saturated salt solution we can also use a very dilute sulphuric acid (0·6 p.c.) to precipitate the B-fibrinogen. This reaction is important, as it shows that the B-fibrinogen cannot be simple paraglobulin. I have made special experiments to convince myself that paraglobulin is dissolved by even a trace of this acid.

The dilute sulphuric acid, however, precipitates the A-fibrinogen from its solutions. In this test we have a means of estimating the richness of various forms of plasma in precursors of fibrin. For instance, to fresh peptone-plasma we add the dilute acid till the reaction is strongly acid. By this means all the precursors of fibrin in the plasma are precipitated. If we collect and wash the precipitate and dissolve it in a small quantity of very dilute alkali, the solution clots spontaneously. If, on the contrary, we precipitate cooled plasma in the same manner, and dissolve the precipitate as before, the solution will now only clot on addition of ferment or leucocytes. The experiment can also be made with sodium chloride plasma or magnesium sulphate plasma.   In the first case, we get a solution which is spontaneously coagulable, owing to the presence of the A-fibrinogen ; in the second, in which this substance is absent, the solution will clot only on addition of ferment or leucocytes. Although it follows from these reactions that neither A- nor B-fibrinogen can be paraglobulin, yet both substances are proteids, as their reactions show.   At the same time, they always contain considerable quantities of lecithin ; whether in combination or as a mixture I cannot at present decide.

The Interaction of A- and B-fibrinogen in the Production of Fibrin

The nature of this reaction is best studied in peptone-plasma. So long as A-fibrinogen is present, this can be made to clot by means of a stream of CO_2.   Before the clotting the plasma is entirely free from fibrin ferment, but after the clotting fibrin ferment is present.   As A-fibrinogen is gradually removed, the plasma clots more slowly with CO_2, and less ferment is formed. If by means of long cooling all the A-fibrinogen be removed, the plasma can no longer be made to clot by a stream of CO_2, and no ferment is formed.

If now to the plasma we add A-fibrinogen (that is to say, so long as this is not too much altered), it will again clot on the passage of a current of CO_2.   In the clotting of the plasma there is a total conversion of its fibrin-factors into fibrin.   However, it is not necessary to add A-fibrinogen itself ; one of its constituents suffices, namely, *lecithin*.   If we add this substance to the cooled plasma, and then conduct through the plasma a stream of CO_2, it clots just as well as if all the A-fibrinogen were present, and in this clotting a large amount of fibrin ferment is formed.

Thus we see there exists a great analogy between the action of lecithin and that of the A-fibrinogen ; and if we remember that the latter contains lecithin, it seems exceedingly probable that the action of the A-fibrinogen depends on the lecithin it contains.   We might conceive that the first effect of the lecithin is to cause the formation of ferment, and that the ferment then reacts on the B-fibrinogen, which has been altered by the stream of CO_2. This, however, is not the case. With regard to this conception, the following experiment is instructive. A peptone-plasma is prepared which is not quite fully peptonised. I must here remark that the behaviour of peptone-plasma differs somewhat, according as more or less peptone than the ordinary dose (0·3 grm. per kilo.) has been injected. If the dose is smaller, the plasma is much more easily made to clot, and we need only add lecithin to bring about the clotting; and the same holds good for normal cold plasma. For example: slightly peptonised plasma is cooled, so that the ordinary precipitate of A-fibrinogen is produced. This assumes very soon a fibrinlike consistence. The clear plasma is now separated into two portions: one half is treated with lecithin and cooled to 0° C.; a large quantity of a coarsely granular precipitate is produced, which soon forms a mass of fibrin, so that the whole plasma becomes solid. This fibrin corresponds in all respects with the ordinary fibrin. The second half is treated with ferment and cooled to 0° C.; no change takes place. I think the experiment may be interpreted by assuming that the B-fibrinogen requires the presence of lecithin for its change into fibrin, and that in this action the ferment plays no part.

The direct proof of an action of the lecithin on the isolated B-fibrinogen is much more difficult to give, and for the following reasons. As I have shown above, B-fibrinogen on its separation from the plasma very quickly undergoes alteration into Hammarsten's fibrinogen. On the latter substance neither lecithin nor lymph-corpuscles have any effect. So long as the alteration of the B-fibrinogen has not gone too far the lecithin promotes coagulation, but it is very hard to control the stage of this alteration, especially in dog's blood. I have only succeeded in horse's blood. For this experiment the blood is allowed to flow from the vein into magnesium sulphate solution (1 volume $MgSO_4$ solution to 3 or 4 of blood). The plasma, which is free from A-fibrinogen, is precipitated by addition of the same volume of NaCl solution. The finely flocculent precipitate is collected by

means of the centrifuge, dried between filter-paper, and then
dissolved in weak alkali. The solution will last some time without clotting, but clots very quickly on addition of lymph-corpuscles. It clots, moreover, but rather slowly, on addition of
lecithin. The experiment is generally successful, though at
times it fails. This is probably owing to the fact that, as we
mentioned above, too free dilution of the solution changes its
properties. Secondly, it is often difficult to diffuse the lecithin
satisfactorily through the fluid on account of its insolubility.

I have prepared the lecithin for these experiments from various sources—brain, testis, lymph-glands, thymus. The lecithin
from all these sources was efficient. On the other hand, it is
possible to prepare a lecithin out of hens' eggs or spawn of fish
which has not the slightest action on plasma. In these cases
we are probably dealing with different kinds of lecithin. With
the exception of their behaviour towards plasma, they all possess
similar properties, especially the formation of a precipitate
with platinum chloride, their richness in phosphorus, the
formation of myelin drops, their solubility in alcohol and ether.
If the solution of lecithin prepared as I have described is
precipitated with platinum chloride, the supernatant fluid, containing the small quantity of impurities, is totally devoid of
any action on plasma, while lecithin recovered from the platinum
precipitate is as efficient as before.

Besides the two bodies which I have denoted as A- and B-
fibrinogen, and which are present in blood-plasma, it is possible
to prepare other substances from various sources which possess
considerable power of bringing about coagulation. These substances are all proteids in combination with a good deal of
lecithin. They can be procured from the testis, lymph-glands,
chyle, brain, thymus, and stroma of the red blood-corpuscles.
They are not perfectly identical with one another, although they
possess many properties in common. It is especially noteworthy
that they are insoluble in very diluted sulphuric acid, or in diluted
acetic acid, a reaction which distinguishes them sharply from
paraglobulin. These substances will not only bring about coagulation in extravascular plasma, such as peptone-plasma, but on

injection into the circulation they cause the most extensive intra-vascular clotting. In this clotting, whether extra- or intra-vascular, the substances disappear as such, and as the quantity of the fibrin found varies with the quantity of the substance injected, we must assume that in part at least it is converted into fibrin. These substances, too, like the lymph-corpuscles and lecithins, are wholly without effect on dilute magnesium sulphate plasma, or on a solution of Hammarsten's fibrinogen. Thus they can contain no fibrin ferment. When, however, they cause extravascular clotting, a large amount of ferment is simul-taneously produced, thus illustrating again the analogy with the action of pure lecithin.

In conclusion, I should like to make some remarks on the fibrin ferment. There can be no doubt that from serum we can prepare a substance possessing all the properties which Alexander Schmidt ascribes to the fibrin ferment, especially the power of changing a certain form of fibrinogen (the fibrinogen of Schmidt and Hammarsten) into fibrin.

The important *rôle*, however, which has been ascribed to this substance in the clotting of normal plasma can no longer be maintained. It has been assumed that the ferment is wanting in the blood which has just left the vessels, and that it is then produced through the death and destruction of the white corpuscles. Disregarding the fact that it is very doubtful whether the white corpuscles have anything to do with this sub-stance, we must at any rate look upon it as certain that the ferment is not produced in so simple a manner. Dead leuco-cytes out of lymph-glands contain no fibrin ferment and have no effect on Hammarsten's fibrinogen, or on very dilute $MgSO_4$ plasma. On the other hand, they produce clotting in various kinds of plasma, and simultaneously fibrin ferment appears, although this is present in neither the cells nor the fluid plasma.

In like manner the fibrinogens which may be procured out of the tissues, thymus, testes, &c., are quite free from ferment, and without effect on Hammarsten's fibrinogen, or on dilute $MgSO_4$ plasma, although they produce clotting in normal plasma, and at the same time give rise to the production of fibrin ferment.   These substances cause intravascular clotting; and it has been already mentioned that the amount of the clot formed increases with the quantity injected.   If the quantity injected is sufficient, the whole fibrin of the blood can be separated out, and in this process only a nominal amount of ferment is formed.   It is evident from this that Hammarsten's fibrinogen cannot exist within the vessels, for this clots only with ferment, *not* with mother substances of ferment, such as leucocytes, tissue-fibrinogen, stroma, or lecithin.

In accordance with this, it is possible to inject even large quantities of fibrin ferment into an animal without causing intravascular clotting.

CLASS VI

FIRST SECTION OF REPORT TO THE SCIENTIFIC COMMITTEE OF THE GROCERS' COMPANY

FIRST SECTION OF REPORT TO THE SCIENTIFIC COMMITTEE OF THE GROCERS' COMPANY[1]

THE NATURE OF COAGULATION

I. *White Corpuscles and Coagulation*

THERE is at the present day a very prevalent opinion that white corpuscles are essentially concerned in the coagulation of shed blood. The only real opposition to this doctrine lies in the substitution of a special kind of form-element, the so-called 'hæmatoblasts' or 'Blutplättchen,' for the white corpuscles. The views held by different observers on the Blutplättchen are extremely diverse. Some deny the participation of white corpuscles in coagulation altogether, and attribute the same powers to the Blutplättchen which Alexander Schmidt has regarded as being possessed by white blood-corpuscles (Bizzozero). Others, whilst attaching great importance to the Blutplättchen in certain forms of intravascular clotting, deny that they have any influence whatever in the coagulation of shed blood (Eberth and Schimmelbusch), and Löwit denies that Blutplättchen are form-elements at all, but affirms that they are globulins precipitated from the plasma (which is readily converted into fibrin by the fibrin ferment), and have no importance in bringing about the process of clotting in shed blood. The introduction of the Blutplättchen into the coagulation question has done nothing whatever to solve the really important biological problem as to whether the plasma clots by itself, or whethe it must have the aid of cells.

In previous publications I have very plainly indicated the

[1] Dated Nov. 4, 1888.

views which I myself have adopted as the result of a very wide investigation on the subject, and in the present section I shall repeat the evidence which has led me to these conclusions, and shall, further, offer some additional facts in support of my opinions. The results which I have previously stated may be shortly summarised as follows : The white corpuscles are not necessary to coagulation ; if they play any part at all, it is a very insignificant one. The character of instability, the rapid power of change in response to changed outward conditions—on which the coagulation of the blood depends—resides not in the corpuscles, but in the plasma of the blood.

My conclusions with regard to the Blutplättchen cannot be so shortly stated.[1] In the first place, no one has yet proved that any such *structures* exist. It is an undoubted fact that disclike bodies, which are neither red nor white corpuscles, can be seen in shed blood and, under certain conditions, in the blood in the vessels ; and these discs are regarded by many observers as anatomical structures. But discs can be obtained from perfectly pure plasma, in the form of a precipitate—discs of which it may be said with certainty that they are not anatomical structures, but a peculiar kind of precipitation analogous to crystallisation. The observers who have maintained the anatomical character of Blutplättchen have not always taken into account the existence of this peculiar precipitate, the microscopical appearance of which is extremely like that of the blood-plates seen in blood ; and, until some characteristic distinction can be madeout between the two, the existence of Blutplättchen must be a matter of doubt. Löwit is disposed to deny altogether that there are such structures : so far as my observations go, I am of the same opinion ; but neither the observations of Löwit nor of myself allow a positive denial to be made.

I do not think that the forms seen and described in blood as Blutplättchen are all of the same nature ; the bulk of them consist of a constituent of the blood-plasma, to which I give the name ' A-fibrinogen.' I regard this substance as of great importance in coagulation, both intravascular and extravascular.

---

[1] The whole controversy upon this point is not attempted in this Section.

I regard it as the agent in initiating coagulation of both kinds, and I differ in this respect from Löwit, who, while agreeing with me that the Blutplättchen are a proteid precipitate, denies its having anything to do with the setting up of coagulation, and considers it merely as a substance readily converted by fibrin ferment and other agencies into fibrin or into a body closely resembling fibrin. Löwit has chiefly investigated the matter from the microscopical side. He adopts the view that coagulation is a process which depends on the action of a ferment upon a proteid, and that, so far as the initiation of coagulation is concerned, the only point to be determined is, where does the ferment come from ? and, according to him, it does not come from the proteid precipitates, which he regards as identical with Blutplättchen. This view of Löwit—which is rather an opinion on his part than a carefully worked-out result—appears to me the opposite of the truth, and I think that the fibrin ferment does owe its origin to this substance ; but I further think that a decision as to the origin of the ferment by no means solves the problem of coagulation, since I hold that the fibrin ferment is utterly unimportant in the process.

One of the chief results of my observations on coagulation is that the plasma itself contains all that is necessary for coagulation, and is so constituted that the phenomena of spontaneous coagulation can be readily explained without its being necessary to invoke the interference of any form-elements.

Apart from microscopical evidence, which, as is well illustrated by the controversy on Blutplättchen, is absolutely contradictory, the reasons for regarding the white blood-corpuscles as being the great agents in initiating coagulation were, until the last few years, afforded by the observations of Alexander Schmidt and his pupils. These observations consist partly in experiments on shed blood and partly in experiments on the intravascular changes of blood when certain substances are injected into the circulation. The experiments on shed blood consist in the main of the following : By allowing blood from the vein of a horse to flow into long vessels surrounded by ice, the coagulation of the blood is greatly delayed and the corpuscles are able to sink.   The red corpuscles sink more readily than the white, the latter partly forming a layer on the surface of the column of red, and partly being scattered through the lower layers of the plasma.   Ultimately, in spite of the cold, coagulation takes place, and it is then observed that clotting begins earlier and is most complete in those strata of the plasma where there are most white corpuscles.   Again, by filtering the plasma at a low temperature, before coagulation has had time to occur, the complete removal of the white corpuscles can be effected, and the plasma so obtained is found to clot very slowly and scantily, even at ordinary temperatures.

The experiments are described in detail in Pflüger's *Archiv*, 1876, and are apparently very convincing.   But their value entirely depends on the signification to be attached to certain granular matter which always accompanies the white corpuscles when they are allowed to settle by subsidence, and which must necessarily be removed by the same process of filtration that separates the leucocytes.   Now Schmidt arbitrarily assumes that this granular matter, which is very abundantly present, consists in fragments of white blood-corpuscles.   The experimental methods then at his disposal did not permit of a satisfactory answer being given to the question as to the nature of these granules.   He makes the observation that the removal of white blood-corpuscles plus granular matter from plasma affects the coagulation of the plasma in two directions: it greatly diminishes the tendency unfiltered plasma has to coagulate spontaneously, and it lessens the quantity of fibrin formed. Now, plainly, these two results might with equal justice be referred to the granular matter as to the leucocytes, and to get out of this difficulty and to fix the responsibility on the white corpuscles, Schmidt assumes that the granular matter is *débris* of corpuscles.   If, by other methods of experimentation, it can be shown that the very process used by Schmidt—*i.e.* cooling down to 0°—is not an indifferent procedure for the plasma, but causes the precipitation from the latter of a proteid substance which can be clearly shown to be of the greatest importance in coagulation, it is obvious that these experiments of Schmidt lose their value as decisive evidence that the white corpuscles play a predominant part in coagulation.

Peptone-plasma is obtained by injecting a solution of peptone into the jugular vein of a dog and subsequently bleeding the animal. By means of the centrifugal machine a perfectly clear plasma is obtained, absolutely free from elements of all kinds. It is convenient to retain the name ' peptone-plasma ' for the plasma thus obtained. Fano, who first investigated the action of peptone on the blood, used a substance which, under the nomenclature in use at the time, could be called by no other name than ' peptone.' In fact, a product was used from which all the known albumoses had been removed by the hydroferro-cyanic acid method ; later a peptone, specially prepared for this purpose by Dr. Grübler, of Leipzig, according, in the main, to Henniger's method, was used. The results in both cases were identical. All the facts which I have published concerning peptone-plasma have reference to a plasma obtained by inject-ing peptone prepared in this way.

Peptone-plasma does not possess absolutely constant pro-perties ; it varies a little according to the condition of the animal—i.e. its state of nutrition, the period of the last meal, the nature of the previous food. Similar variations occur in all kinds of extravascular plasma, and the blood in the vessel varies under like conditions. These facts will be discussed in greater detail in another section. At present, in speaking of peptone-plasma, I refer to plasma which is most commonly obtained, noting any special modification as occasion may require. Pep-tone-plasma, free from all form-elements, is spontaneously coagulable, for it coagulates when a current of CO_2 is passed through it, or when it is neutralised by other acids—e.g. acetic acid. It also clots on dilution with water, or, though much less readily, with $\frac{1}{2}$ per cent. NaCl solution. Further, filtration through a clay cell will cause coagulation to a certain extent. These processes cannot be regarded as ' fibrin factors ' in the ordinary sense of the word. The natural process of coagulation has been inhibited by the peptone injection, and by the above-mentioned means this inhibitory effect is removed.

Fano was of opinion that the peptone injection influenced the white blood-corpuscles. The views of Alexander Schmidt were at this time in full force, and I accepted Fano's conclusions in my earlier experiments, the more so as my own observations on the leucocytes from lymph-glands pointed strongly in the direction. Fano had occasionally seen that continued centrifugalising rendered the plasma less readily coagulable by the above-mentioned means, a result which I repeatedly observed. The coagulation which was induced by the CO_2 was supposed to be due to its action on the white corpuscles which had escaped removal. Subsequently, I found that this conclusion, or, more strictly, supposition, was not correct.[1] The prolonged centrifugalising may have entirely removed the white corpuscles, but it certainly removed something else ; for I found that the action of the centrifuge was without any influence on the plasma, unless the latter was kept during the intervals in a cool place. I found, moreover, that plasma—which had been kept at ordinary temperatures and had, over and over again, been subject to the centrifuge until absolutely no precipitate was produced, which was perfectly clear, and in which the most careful and repeated examinations failed to discover any trace of form-elements— became rapidly turbid when it was cooled down to 0°, and that a granular precipitate was deposited.

Peptone-plasma, perfectly free from form-elements, is spontaneously coagulable by a current of carbonic acid, &c.; but this spontaneous coagulability is lost if the plasma be subjected to prolonged cooling to 0°, and the precipitate which is so formed be removed. The addition of this precipitate—certain precautions being adopted—again renders the plasma spontaneously coagulable. Now, it may be fairly concluded from the results obtained with peptone-plasma, that it is incorrect to assume, as Alexander Schmidt does, that cooling plasma down to 0° is an indifferent procedure, or that the large quantities of granular matter found in this plasma are necessarily broken-up white corpuscles. It may, of course, be said that peptone-plasma is

---

[1] Ueber einen neuen Stoff des Blutplasmas ' (*Du Bois' Archiv*, 1884) contains the first mention of this ' cold body.' [Collected Papers, p. 124.]

not a normal plasma. This objection, in fact, is often made. It must, however, be remembered that we have no criterion whatever as to what is a 'normal extravascular plasma.' In all the experiments which were made in the early history of coagulation two main methods were used. In the first, the blood was prevented by artificial means from clotting, as it normally would do. The mode of action of the method adopted has been explained by arbitrary assumptions lacking all real foundation, not the slightest information having been acquired as to what influence the method adopted might exert on the plasma. A careful study of the methods used in investigating coagulation will show us that all the various means adopted to prevent clotting give plasmata which are markedly different in different cases, and that, further, every kind of extravascular plasma differs from the plasma of the circulating blood. In the second method, the various transudation fluids have been used as identical with normal plasma; but it can be shown, in the most striking manner, that these transudation fluids are profoundly different from blood-plasma, and hence results obtained from the use of these fluids have not of necessity any bearing on the nature of the normal coagulation of the blood, since the assumption that the transudation fluid is identical with the blood-plasma is not true.[1]

The study of peptone-plasma leads to the conclusion that cooling, the method used by Alexander Schmidt to prevent coagulation, has a pronounced influence on the plasma, inasmuch as it removes from the plasma a substance which is of great importance in initiating coagulation. Owing to technical difficulties, which it is at present impossible to overcome, it cannot be stated that the granular matter described by Schmidt is identical with the precipitate produced by cold in peptone-plasma. Very valuable confirmatory evidence is, however, afforded by observations on salt-plasma. For our present purpose we need only consider two well-known kinds of salt-plasma. The first variety we may call 'weak salt-plasma.' It is obtained by receiving blood direct from the artery of a dog into an equal volume of 8 or 10 per cent. solution of common salt.

[1] Cf. later portions of this Section in this connection.

The plasma is obtained by the centrifuge. On diluting this plasma with four or five times its volume of water it clots spontaneously. The rapidity varies. It may take from two to thirty minutes—the difference, as I shall afterwards point out, depending probably on the state of the animal's nutrition. It is always spontaneously coagulable, and it retains this power in spite of long cooling. The second variety is obtained by receiving blood direct from the vessel into a solution of sulphate of magnesia. The proportion of saturated solution of sulphate of magnesia to blood should be as 1 to 3 or 4. It is preferable, in the case of the dog, to take a half-saturated solution of sulphate of magnesia and an equal volume of blood, or to two volumes of half-saturated $MgSO_4$ solution three volumes of blood. In dealing with dog's blood, it is of very great importance that the slightest possible delay should occur between the blood leaving the vessels and its thorough admixture with the salt solution. This is one reason for using the weaker solution, since the admixture takes place more rapidly; another reason is that, with this strength of $MgSO_4$, the red corpuscles are less liable to be destroyed, and it is essential to obtain a plasma absolutely free from form-elements and from the products of the destruction of the red corpuscles. This strong salt-plasma (as we will call it) does not coagulate spontaneously on dilution, though to this general statement there is a very important exception which we shall afterwards discuss.

These two varieties of salt-plasma, one of which is spontaneously coagulable and the other not, have long been known, and their different behaviour on dilution explained by arbitrary assumptions, viz. that the plasma required some addition from the corpuscles in the shape of fibrin ferment, when it was further supposed that, whereas the weak salt-plasma allowed the ferment to pass out, the strong salt-plasma prevented its doing so; and the non-coagulation of the weak plasma until diluted was imagined to be due to the ferment being unable to act on account of the salt. But, as a matter of fact, the fibrin ferment [1]

---

[1] 'On the Origin of the Fibrin Ferment,' *Proc. Roy. Soc.* 1884. [Collected Papers, p. 119.]

is not present in the weak salt-plasma any more than it is in the strong, and the coagulation of weak salt-plasma on dilution is altogether inexplicable on the old grounds. Whether the different salt solutions act differently on white blood-corpuscles or not is a matter of pure conjecture, as we have absolutely no means of definitely ascertaining this point; but there are the strongest reasons for thinking that the plasma is very materially altered by different salt solutions. The evidence as to the alteration of the plasma is as follows: 1. Peptone-plasma free from form-elements is obtained, and it is essential for these experiments that a plasma readily coagulable on dilution should be used. To this plasma, saturated solution of sulphate of magnesia is added in quantity just insufficient to produce an immediate precipitate. One portion of this plasma is diluted with nine or ten times its volume of water: it coagulates spontaneously in fifteen to twenty minutes. The other portion is allowed to stand, and after a few hours a marked precipitate is formed. The precipitate is removed, and the plasma, when now diluted, does not clot on standing for twenty-four hours, but clots readily on addition of fibrin ferment. 2. Peptone-plasma is mixed with an equal volume of 10 per cent. NaCl: it clots readily, on dilution, after five minutes, and equally well, when diluted, after three days in the ice-chest, no appreciable precipitate being formed in spite of the cold. The above experiments can be constantly and readily repeated, provided the precautions as to peptonisation, which will be described later on, are observed. The following is exceptional: 3. Dog's blood is received into strong sulphate of magnesia and centrifugalised, the perfectly clear plasma diluted immediately, when coagulation occurs in ten to fifteen minutes. The plasma is allowed to stand overnight and a coagulum-like precipitate is formed; the plasma is now diluted, but it has lost all power of spontaneous coagulation. I have observed this result in at least twelve experiments, but it cannot be obtained with certainty. The success depends on the correct proportion of salt and blood. A weak solution—i.e. 10 per cent. $MgSO_4$ and an equal volume of blood—generally yields a spontaneously coagulable plasma: a very strong solution

P

—*i.e.* one volume $MgSO_4$ to three of blood—always gives a plasma not spontaneously coagulable on dilution. A successful result can be obtained with a proportion of four or five volumes of blood to one volume of saturated $MgSO_4$, but it is not invariable. The deposit of a fibrinous precipitate is frequent, but it is often so small in amount that it will easily be understood, from a subsequent section of this work, why no spontaneous coagulation occurs on dilution.

The older experimental literature on extravascular blood does not give any information of the slightest use as to whether the white corpuscles are concerned at all in coagulation. On the other hand, it is apparent, from what I have adduced above, that the plasma [1] is spontaneously coagulable, and that its spontaneous coagulability is due to the presence of a substance in solution which can be removed by cooling, or by adding sulphate of magnesia of a certain strength. In 1881 I published, in the ' Proceedings of the Royal Society,' observations which throw a quite different light on this subject. Using isolated leucocytes from lymph-glands, I found that these, added to peptone-plasma, very readily produced coagulation. This experiment, which can readily be repeated, was speedily confirmed by Rauschenbach,[2] using cooled plasma; and it is apparently strong evidence that white corpuscles are concerned in coagulation. But it is only evidence if we assume, as many do, that white blood-corpuscles and lymph-corpuscles from lymphatic glands are identical structures. It is not to be doubted that white corpuscles are derived from lymph-corpuscles. The point to be considered is whether the passage of a leucocyte into circulating blood-plasma is without any influence on the corpuscle. If leucocytes from lymph-glands be added to extravascular plasma, the result is not indifferent either for the lymph-cells or for the plasma, so that we have *a priori* no right to suppose that such is the case when leucocytes reach the plasma in the ordinary course through the lymphatic current.

Let us first examine the action of leucocytes on extravas-

---

[1] *I.e.* certain forms of extravascular plasma.

[2] Rauschenbach, Dissertation, Dorpat, 1883.

cular plasma, using for the purpose peptone-plasma. As I have mentioned above, and shall explain more fully subsequently, peptone-plasma varies, and to a large extent these variations can be determined at will. The general result of adding leucocytes to peptone-plasma is to cause clotting. The result is, however, not invariable, and the process, when studied more exactly, is seen to vary very considerably in different examples of plasma. I will illustrate this by the following experiment. Two equal portions of peptone-plasma are taken. No. 1 is 'weak' peptone-plasma; No. 2 is 'strong' peptone-plasma.[1] To both, isolated leucocytes, suspended in $\frac{1}{2}$ per cent. salt solution, are added in the proportion of half a volume of leucocytes to one volume of plasma. Clotting occurs in both. They are allowed to stand twenty minutes, when the clot is squeezed out and the serum collected. To the serum a similar further addition of leucocytes is made. Serum No. 1 does not clot; serum No. 2 clots through and through so as to form a solid mass. Clot No. 2 can again be squeezed, and the serum again clots on addition of leucocytes. The experiment can be altered so that, to both specimens of plasma, a small quantity of leucocytes be added. Plasma No. 1 will clot completely, so as to form a solid mass; plasma No. 2 will clot, but the clot—as can be seen by adding numerous small and varying quantities of leucocytes—is roughly proportional to the cells. How closely proportional the amount of clot and that of the added leucocytes is may be gathered from the following quantitative determination. A large quantity of washed leucocytes were suspended in $\frac{1}{2}$ per cent. salt solution. The mixture was well shaken so as to produce as uniform a distribution of the cells through the fluid as possible. The fluid was then divided into two equal portions: one was added to 20 c.c. plasma ('strong'), the other was used to determine the weight of cells added. The weight of clot produced was 0·21 grm., the weight of cells added was 0·17 grm. As a control to this, the weight of proteid present in the serum before and after coagulation—allowance being made for the dilution—was determined, and found to be 6·26 grms. per cent. before and

---

[1] [The explanation of these terms is given on pp. 195 and 232.]

6·30 grms. after coagulation with cells. Taking into consideration what I have said above concerning the repeated and limited
clottings, and the fact that quantitative determinations like
those given above are necessarily not free from slight errors, it
would appear that, in the case of the coagulation occurring in
'strong' peptone-plasma, the clot is chiefly if not entirely
formed by the cells. It is, indeed, difficult, from observations
on strong plasma, to be sure that the plasma itself is involved in
the coagulation, that is, as regards its yielding a part of the
material substratum of the clot. But that it is so involved is
extremely probable from observations which can readily be made
on weak peptone-plasma. Here crop after crop of coagulum
corresponding in the main to the quantity of cells added, cannot
be obtained. A single addition of leucocytes, in small quantity,
suffices to produce a complete clotting of the plasma, and to
deprive it of the power of clotting when more leucocytes are
added, provided time enough be allowed to elapse. It is readily
possible to observe this result in the case of weak peptone-
plasma : a small quantity of leucocytes is added, and as soon as
a clot has made its appearance it is removed, and the fluid
gently squeezed out. The leucocytes are contained in the clot.
Nevertheless the fluid clots through and through, there being
obviously no direct relationship between the amount of cells
added and the clot formed. A further instance of the cells'
influence on the plasma is seen in the fact that, after the clotting with the cells, the plasma has lost the power of coagulating
when a stream of carbonic acid is passed through it.

It must be distinctly understood that, when leucocytes are
added to plasma, there is no question of a destruction, in the
sense of a disappearance of the cells, taking place. If the clot
be examined under the microscope, the protoplasm of, at any
rate, the bulk of the cells has become merged in the fibro-
granular ground-substance, although the nuclei are entirely
unaltered, so far as can be made out.

But, although the general result of adding leucocytes to
peptone-plasma is to induce clotting, there are exceptions. In
very strongly peptonised plasma the leucocytes will not cause

clotting, and they are apparently unaltered so far as can be judged by microscopical observation; but they are profoundly altered physiologically, for if they be collected by the centrifuge and added to a portion of another plasma, which clots very easily with leucocytes, they have no effect; that is to say, their sojourn in the strong peptone-plasma has deprived them of the power of setting up clotting in weak plasma.

Finally, when leucocytes have been added to peptone-plasma until its power of clotting with cells is exhausted, the plasma has always lost the power of clotting by CO_2, by dilution, or by other means which will normally make the plasma clot. In fact, it would appear that the fibrinogen of the plasma had been used up. This is, however, not so : in the case of weak plasma minute quantities only are left; in the case of strong plasma the amount is, as far as can be made out, unaltered.[1]

The conclusions to be drawn from the observations on the interaction of leucocytes from lymph-glands and extravascular peptone-plasma are as follows. If plasma and leucocytes come together, both are affected : clotting may ensue in which both factors are *materially* concerned ; clotting may ensue in which only one of the factors is materially concerned, the other not being appreciably diminished, but having had its power of clotting destroyed or interfered with ; no clotting may ensue, but one of the factors may lose its power of inducing clotting under other circumstances. In other words, we have, in dealing with extravascular plasma, two distinct phases to note : a positive clotting phase, and a negative, inhibitory to clotting, phase ; and further, as regards extravascular plasma— a variable product—sometimes the positive, sometimes the negative, phase may predominate. With this knowledge, the action of leucocytes on circulating blood, and the conclusions which may be deduced as to white blood-corpuscles, may be more advantageously considered.

The first experiments made on this subject were recorded in my paper in the ' Proceedings of the Royal Society ' in 1881, and in a paper in Du Bois Reymond's ' Archiv' for 1881. The

---

[1] This will be discussed in greater detail in later Sections.

experiments consisted in injecting washed leucocytes into the circulation. The leucocytes were obtained from fresh lymphatic glands, were well washed with ½ per cent. NaCl solution, and were injected in large quantity into the jugular vein of the dog. No effect was produced, either when the dog was in its normal state or after it had been injected with peptone. The quantity injected was very large—25 to 30 c.c. of a stiff admixture of cells into dogs of 7 to 5 kilos. in weight. An experiment was also made in which the cells were injected into an isolated vein. In no case was there the slightest trace of intravascular clotting; although the cells, when added to extravascular blood in small quantity, immediately produced clotting. A very important conclusion could be drawn from these experiments, viz. that the impetus to coagulation when blood leaves the vascular wall lies, not in the death of leucocytes, as Alexander Schmidt concluded, but in a change of the blood-plasma, seeing that extravascular plasma clots readily with dead isolated leucocytes, while intravascular plasma will not so clot.

The researches were repeated some years later by Dr. Groth[1] in A. Schmidt's laboratory, the results obtained being directly opposite to my own. Dr. Groth did not, however, use leucocytes, but the expressed juice of lymphatic glands, the injection of which caused intravascular clotting in many instances. The curious explanation Dr. Groth gives of this intravascular clotting will be discussed later on. Following the established tradition of the Dorpat school, Dr. Groth assumes that, when he injects leucocytes plus the fluid in which they occur, the result must necessarily be due to the leucocytes and not to the fluid. It has always been the custom of Alexander Schmidt and his pupils, when two factors are concerned in any given result, to assume that the one they arbitrarily choose must be right. Dr. Groth in this case chooses the leucocytes as the agents in producing intravascular clotting. Shortly after the paper of Dr. Groth appeared I had succeeded in isolating from the testis and thymus

---

[1] Dr. Otto Groth. 'Ueber die Schicksale der farblosen Elemente im kreisenden Blute.' Dorpat, 1884.

a substance which, when dissolved in alkaline $\frac{1}{2}$ per cent. NaCl solution, invariably caused intravascular clotting; and I found that this substance was present in the fluid of lymphatic glands. I repeated my previous experiments with isolated leucocytes, and found that they did not produce intravascular clotting. Subsequently, Dr. Krüger,[1] assistant to Professor A. Schmidt, again takes up the question, and endeavours to prove that leucocytes do cause intravascular clotting. The evidence which he adduces in his paper in support of the fact that it is the leucocytes and not the intercellular fluid which produce the intravascular clotting is as follows. He first of all tries to show that the intercellular fluid of lymph-glands is wholly without influence when injected into an animal, but his conclusions are fallacious. He chops up lymphatic glands and squeezes out the juice. This is a grumous, sticky, slimy mess, on which the centrifuge has practically no influence : however, he lets the centrifuge separate as much as it can of the cells. The supernatant sticky fluid he filters through several layers of thick filter-paper. The filtrate, injected into animals, is without action ; the residue, consisting both of cells and fluid, when injected, causes intravascular clotting; consequently, he considers that the fluid of lymphatic glands is without influence. This conclusion is erroneous. The fluid of lymphatic glands contains a large quantity of a substance which I have spoken of as lymph-fibrinogen.[2] It is pre-existent in the fluid of these glands, and is closely allied to the tissue-fibrinogen afterwards described. The fibrinogen of lymph-glands, just as the fibrinogen of blood, is not in a state of solution, but in a semi-dissolved swollen-up condition, and it filters with great difficulty, so that, should any solid bodies be present to stop up the pores of the filter, it does not filter through at all. If to the fluid of lymphatic glands acetic acid be added, a dense precipitate occurs, and microscopical examination shows that this precipitate is formed in the intermediate fluid. If the fluid of lymphatic glands be diluted

---

[1] *Zeitschrift fur Biologie*, vol. xxiv.

[2] 'Croonian Lecture Abstract,' *Proc. Roy. Soc.* 1886. [Collected Papers, p. 157.]

with normal salt solution, and centrifugalised until all form-elements be removed, it gives an abundant precipitate with acetic acid ; but should it be filtered through thick filter-paper, only the first small portion of the filtrate gives any pre-cipitate with acetic acid.

It is this acetic acid precipitate which is the active agent in the fluid of lymph-glands, and if the fluid or the acetic acid precipitate dissolved in dilute alkali be injected into an animal, it infallibly causes death.   This behaviour of the fibrinogen of lymph is quite characteristic of all the group, as they occur in nature.   The fibrinogen of blood will not filter through a clay cell if the plasma in which it is contained be a little fatty. Had Dr. Krüger examined his filtrate, he would have found that it contained only traces of proteid and no fibrinogen ; and I think no one would be disposed to admit that chyle and lymph are practically free from proteid and fibrinogen.

With regard to the action of isolated cells, although the paper of Krüger is ostensibly written to prove that they are extremely active in producing intravascular clotting, no single experiment is recorded in which washed leucocytes were injected. In one experiment the fluid was mixed with a certain quantity of salt solution, and the centrifugal precipitate injected, when it caused thrombosis.   But no attempt was made to ascertain whether the intermediate fluid had really been removed by the centrifuge (which, indeed, never happens with a single centri-fugalisation), and from the context it is obvious that, in this particular case, a large quantity was present.   In the next experiment the cells were washed with minute quantities of water and NaCl solution respectively.   There is no guarantee whatever that the cells were effectually washed, and the appli-cation of distilled water entirely vitiates the experiment, for it not only partly precipitates the fibrinogen, but profoundly alters the cells.   Two experiments are described in which intravascular thrombosis is caused in dogs by the injection of ' Zellenbrei,' but it is not stated whether the cells are isolated, or whether merely the cell-mixture of lymphatic glands is injected.   Since no one in Dorpat has ever injected isolated leucocytes from lymph-glands, it is difficult to understand how they can be sure that they always cause intravascular clotting.

Krüger thinks I may possibly be correct when I say that large quantities of leucocytes may be injected without causing intravascular clotting, but imagines that the washing with normal salt solution might remove from the cells the substance efficacious in causing clotting. It is, of course, possible that normal salt solution is not a wholly indifferent fluid for leucocytes. With our present knowledge it is the best fluid to use, and it is certain that we cannot ascertain the action of leucocytes unless we remove as completely as we can the intervening fluid. Dr. Krüger overlooks the real point of these injection experiments. The cells, after thorough washing, are most efficacious in causing the coagulation of extravascular plasma, but they have no power on intravascular plasma to produce clotting; nay, the more the animal be flooded with them the more slowly does the shed blood clot (as we shall see directly), showing that the great effect of bringing leucocytes into circulating blood is to deprive them of the power they possessed previously of causing clotting in extravascular plasma.[1] If the experiments of Groth and Krüger were correct, but one

[1] In these experiments on the injection of leucocytes I make one assumption, viz. that some of these leucocytes must be present in the shed blood. It is a reasonable assumption, because, when we treat leucocytes with extravascular plasma, and therefore know exactly what we are doing, there is no question of a destruction of leucocytes in the sense of their becoming no longer visible, although they are profoundly altered. Groth explains the action of his supposed leucocyte injections by saying that the injected cells and the normal white blood-corpuscles are rapidly destroyed. It is, of course, extremely difficult to understand any mechanism in virtue of which the injection of leucocytes should cause the destruction of white blood-corpuscles, and the method of counting the white blood-corpuscles, before and after the injection, is open to such a very patent objection that it cannot be relied on at all. In Groth's experiments the injection, owing to the admixture of the lymph-fibrinogen, caused very widespread thrombosis. In this case it would be manifestly quite impossible to say how many leucocytes were shut up in the extensive clot, and it is no more permissible to suppose they have been destroyed than it would be to assume that, when normal blood clots and a clear serum exudes, the red corpuscles are essentially concerned in coagulation. There are other objections to this method which I shall subsequently allude to.

inference could be made, viz. that the cells they inject are radically different from the white corpuscles, since quite small quantities of their cells cause intravascular clotting, whereas it is certain that the number of white blood-corpuscles varies greatly without intravascular clotting ensuing.

I will now describe an experiment which is typical of several that I have carried out. It was made with leucocytes from the thymus gland of the calf, but the same experiments have been made with leucocytes from lymph-glands. The thymus was minced and squeezed out in a solution of 0·6 per cent. NaCl. The total quantity of fluid was sufficient to fill four centrifuge tubes. When the fluid was centrifugalised, the cells sank to the bottom. The supernatant fluid (fluid A) was centrifugalised till no further deposit took place. It contained no leucocytes. These latter were again twice washed with 0·6 per cent. NaCl, and the cell deposit, after strong centrifugalising, reached 1½ inch upwards in the tube (the diameter of the tube is about 1 inch). The cell deposit of all four tubes was suspended in 0·6 per cent. salt solution, the total of fluid being 80 c.c. A few drops of this fluid produced rapid clotting in 10 c.c. peptone-plasma. 75 c.c. of suspended cells were injected into the jugular vein of a dog whose weight was 7 kilos. No obvious effect whatever was produced. After about ten minutes the animal was bled from the carotid. The blood flowed in a normally full stream. The animal was killed by chloroform. The most careful search showed an entire absence of intravascular clotting. The blood drawn from the carotid and the blood taken from various parts of the vascular system remained quite fluid after removal from the body. When examined microscopically, it showed numerous leucocytes, which were very unequally distributed, occurring now isolated, now in groups and masses. The blood collected from the carotid remained fluid for two hours (the time it was under observation). But when to a portion of this blood the few remaining leucocytes, *which had not been injected*, were added, clotting took place in two minutes. On the other hand, 20 c.c. of the fluid A injected into a dog of 5 kilos. caused instant death; and 30 c.c. into a dog of 7½ kilos. produced complete clotting of the portal venous system, with a slight clot in the pulmonary artery. I shall leave the discussion of this fluid till later.

In my hands the injection of leucocytes into the circulation of the dog only produces one result—slowing of the clotting of the shed blood. This result, which is only marked if a very large quantity of leucocytes are used, is worthy of note. We have seen in our study of the interaction of leucocytes and extravascular plasma that two processes take place : 1, coagulative ; and 2, inhibitory to coagulation ; that these processes affect both plasma and leucocytes, and that, in extravascular plasma, first one and then the other phase predominates. In intravascular blood there is only one process—that inhibitory to coagulation. When the leucocytes come into the circulating blood they are altered, just as they are when they come into very strong peptone-plasma ; they lose the power of initiating coagulation, but they at the same time change the plasma in such a way that the normal process of coagulation does not occur.

It is certain, from my own observations and those of Rauschenbach and others, that when perfectly fresh leucocytes or other tissue-cells are brought into extravascular plasma, clotting ensues ; and it is equally certain that when leucocytes pass from the lymph into the blood under natural conditions, no clotting ensues, since intravascular clotting is not a normal process. When leucocytes are injected they lose the power they previously possessed of causing clotting in extravascular plasma, since the shed blood, after injection, remains fluid, but clots directly the leucocytes are added. It is, therefore, reasonable to suppose that the same thing takes place when leucocytes normally pass into the blood-current, and that hence white blood-corpuscles are not active agents in producing coagulation when the blood is shed. There are no experiments recorded which show that the injection of leucocytes causes intravascular clotting. It might possibly be achieved if enough were obtained, but I cannot perceive that such a result would throw any light on the participation of white blood-corpuscles in normal coagulation.

My conclusions in regard to the white blood-corpuscles are that—

1. The older experimental evidence is open to the gravest suspicion.

2. Direct microscopical investigation yields results which are so contradictory as to be useless.

3. The apparent evidence which is afforded by the fact that leucocytes from lymph-glands, when added to extravascular plasma, cause coagulation, is only of value if we make the supposition that the circulating blood-plasma is without any influence on the leucocytes. Such a supposition is unwarrantable on two grounds—1st, the experiments in which leucocytes are injected; 2nd, the remarkable provision of nature for preventing blood-plasma from coming into contact with tissue-cells, and the fact that if such contact does take place the tissue-cells are profoundly altered. The plasma which exudes from the vessels is very different from the plasma in the vessels. The former does not interact with tissue-cells.[1] If by experimental means we alter the normal exudation conditions, the cells touched by the plasma are profoundly changed.[2] To the first of these grounds it might be objected that the washed leucocytes from lymphatic glands are not identical with the living leucocytes. Such an objection is largely answered by the second ground.

It is certain, then, that plasma alone, free from elements, has the power of spontaneous coagulation, and all the accurate knowledge we possess points to the non-participation of the white corpuscles. The importance of the question lies in the fact that coagulation is a problem bristling with difficulties. There is a strong prejudice in favour of the white corpuscles being concerned, and a great tendency to refer every new difficulty to the action of the white blood-corpuscles. They are, in fact, so convenient as a spurious explanation of many curious results, that very inadequate evidence of their participation is

---

[1] Vide 'Croonian Lecture Abstract,' Proc. Roy. Soc. 1886. 'Beiträge zur Frage der Gerinnung,' Du Bois' Archiv, 1888. [Collected Papers, pp. 141, 253.]

[2] Vide 'Hæmorrhagic Infarcts of Liver,' Proc. Path. Society, and subsequent sections of Report. [Collected Papers, p. 346.]

eagerly accepted by many physiologists. In my opinion, the adoption of this position is very unfortunate. The study of coagulation has a much wider bearing than the narrow question as to why a particular proteid becomes solid. If we are compelled to introduce white corpuscles into the matter, it will inevitably be found that we explain the bulk of the phenomena by the ' vital activity of leucocytes.'

II. *On Fibrinogen Interaction.   The Negative and Positive Phase of Coagulation*

It has always been held that coagulation is a fermentative process—*i.e.* a process in which a particular enzyme is supposed to act on a particular proteid and alter it in some manner, so that its conditions and solubility are altered. Beyond this, except in the matter of speculation, practically no explanation has been attempted; that is to say, we are wholly ignorant of the nature of the change which the ferment exerts on the proteid. So far as my own observations permit me to judge, I am decidedly of opinion that there is such a body as the fibrin ferment, a body as definite and constant in its action as any of the well-known enzymes; and that this body will convert certain fibrinogen solutions into fibrin. But, for reasons which I shall subsequently adduce, I do not think this body can play any important part in ordinary extravascular coagulation; and I am certain that the importance which has been attributed to it has been such as to entirely obscure the very remarkable phenomena which may be observed in the study of coagulation.

The term ' fibrin ferment ' has been used by many writers in a loose manner, any addition to a coagulable fluid which hastens the coagulation of the latter being spoken of as a ferment. It is, therefore, advisable that any writer should define as nearly as possible what he understands by the fibrin ferment, otherwise great confusion must of necessity prevail. I have always used the term in the sense it is employed by its discoverer, Alexander Schmidt, and I always prepare it according to the method by which Alexander Schmidt prepares it, and test for it in the same way.

The fibrin ferment is a substance contained in blood-serum, and is apparently very little altered by standing for a very long period—weeks and months—under alcohol. From the dried coagulum of serum it is easily extracted by water; the activity of the solutions is destroyed by boiling, and they contain such very minute quantities of organic matter that it is quite impossible to suppose that the latter materially enters into the formation of fibrin.

The test for ferment used by Alexander Schmidt is strong magnesium sulphate plasma (from the horse), diluted eight or nine times. This fluid is not spontaneously coagulable, and is, so far as is known, not coagulated by anything except by ferment. The test for ferment used by Hammarsten is a solution of fibrinogen, prepared from strong magnesium sulphate plasma of the horse by means of repeated precipitation and re-solutions. I think there is no doubt that both Schmidt and Hammarsten refer to the same substance, and when I speak of fibrin ferment I use the word in the sense they use it. The test solution employed by me is generally the strong magnesium sulphate plasma of dog's blood, which acts precisely as that of the horse, and is never [1] spontaneously coagulable on dilution, and clots, so far as I know, only with fibrin ferment. As I have clearly pointed out in my ' Uebersicht einer Theorie,' [2] I regard the normal coagulation of shed blood as due to the interaction of two fibrinogens contained in the plasma of shed blood. It is, therefore, of great importance that I should clearly point out what I mean by fibrinogen interaction, and how essentially this process differs from what is usually spoken of as a fermentative process.

In discussing the reasons which have led me to think that it is improbable that white blood-corpuscles are materially concerned in coagulation, I called attention to a very striking fact, a fact of which, from repeated observation, I am absolutely

---

[1] Never, when used the next day after obtaining it (*v. supra*).

[2] ' Beiträge zur Physiologie,' Festschrift für Carl Ludwig, Leipzig, 1887. [Collected Papers, p. 186.]

certain. When washed and isolated leucocytes are added to extravascular plasma obtained in certain ways, coagulation ensues. Frequently the amount of coagulation is proportional to the amount of cells. On the other hand, when leucocytes are injected into the circulation, a precisely opposite result ensues; the blood is deprived of its power of clotting proportionately to the amount of leucocytes injected. Consequently I say that in the interaction of leucocytes and blood-plasma there are two phases—a clotting phase and a negative to clotting phase.

Now this action can be much more easily studied by using substances which I term 'fibrinogens,' and which can be obtained from blood-plasma, from blood-serum, and from a great variety of tissues. In all probability these substances form the bulk of the protoplasm of cells.

The curious interaction of these substances, and the contrast it forms to the action of the fibrin ferment, has been fully illustrated by me in a paper read before the Royal Society in 1887,[1] and I will shortly repeat the results there recorded.

By rendering ordinary dog's or sheep's serum decidedly acid with dilute acetic or dilute sulphuric acid, a precipitate is obtained. It is not produced until the reaction is decidedly acid. Chemically this precipitate is a proteid, and it contains in addition lecithin. In several trials I was unable to obtain this substance from the serum of horses' blood or from that of bullocks. The following experiments illustrate the action of this substance. Two portions of peptone-plasma were taken. To I. an equal quantity of sheep's serum was added. After many hours it presented a scarcely perceptible trace of clotting. To II. a small quantity of this substance, serum-fibrinogen, obtained from sheep's serum, was added. The mixture became completely solid in fifteen minutes; on squeezing out the clot, the fluid showed no disposition to clot spontaneously, but it clotted directly on adding more of the solution of serum-fibrinogen. Now, to understand this experiment, I must add that the peptone-plasma does not clot with fibrin ferment, that serum contains

[1] 'On a New Constituent of Serum.' [Collected Papers, p. 169.]

only traces of this serum-fibrinogen, and it must be further observed that the clotting goes in stages. The serum-fibrinogen, when added, causes a certain amount of clotting ; then it stops indefinitely; on adding more, clotting again ensues. Hence there is a distinct quantitative relationship between the serum-fibrinogen used and the fibrin produced ; and either the serum-fibrinogen or some part of it is used up in the process. Hence the serum-fibrinogen cannot be called a ferment, if we use the term in its ordinary application.

If a solution of serum-fibrinogen be injected into a rabbit it produces one very marked effect. The shed blood does not clot for several hours. If, however, some of the solution be added to the shed blood, clotting at once ensues, showing that the injected substance must have been altered or used up ; consequently its intracorporeal action is not that of a ferment. The injection of the solution never gave rise to any obvious intravascular clotting. Once a very doubtful trace was found, but in all my other experiments the most careful search showed the absence of all intravascular clotting. Here, then, we have a substance which, like the leucocytes from lymph-glands, produces two contrary results : clotting when added to extra-vascular blood, prevention of clotting when added to the circu-lating blood.

It will now be advantageous to study the action of substances called ' tissue-fibrinogens '—for the present confining ourselves to that of the substance obtained from the thymus, which is, on the whole, representative, and is easily procured. The organ is chopped up and mixed [1] with chloroform water—i.e. water in which a little chloroform is dissolved. It is left to stand twenty-four to thirty hours. The fluid is then centrifugalised, and the supernatant fluid strongly acidified with acetic acid. A floc-culent precipitate is obtained, which is collected and washed by the centrifuge. This precipitate is dissolved in $\frac{1}{2}$ per cent. NaCl solution, to which a few drops of sodium carbonate solu-tion have been added. The solution is filtered. It filters easily,

---

[1] 'On Intravascular Clotting,' *Proc. Roy. Soc.* 1886. ' Ueber intravas. Gerinnung,' *Du Bois' Archiv,* 1886. [Collected Papers, p. 135.]

and gives a clear solution if the precipitate be quite fresh and the extraction have not lasted longer than the time mentioned above. It is a limpid solution, with a very faint opalescence.

The relation of this fluid to the processes of coagulation is now to be considered.

Into the jugular vein of a rabbit some of this solution is injected. As much of the fluid is allowed to run out of the burette as there is time for. Within a time varying from a few seconds to one and a half minute the animal is dead. This result is absolutely constant. If the animal be examined directly, it will be found that the whole vascular system is thrombosed. The only exception is when the death is extremely rapid, in which case the clotting may be confined to the right heart and pulmonary artery. Obviously, in these cases, the clotting is thus limited because the fluid has not had time to reach the general circulation.

If some of this solution be added to peptone-plasma, clotting invariably ensues. If blood be allowed to run direct from the blood-vessel into the solution, clotting occurs almost instantaneously. The following example will illustrate this: Blood was taken direct from the carotid of the dog into two tubes. One tube contained 10 c.c. of fibrinogen solution, the other 10 c.c. of $\frac{1}{2}$ per cent. NaCl; 40 c.c. blood taken into each tube. In the fibrinogen solution the blood was quite solid within fifty seconds; in the other, the blood did not clot at all for six minutes, and it was from seven to eight before it was quite solid. The normal clotting of the blood in this case was exceptionally slow, but the very marked power the fibrinogen has of causing increase in the rapidity of coagulation of the normal shed blood is always apparent and very pronounced. It is, therefore, obvious, from the above statements, that the fibrinogen solution has something to do with clotting, and a more detailed study will show that its action in the matter is of a totally distinct character to any fermentative process. This can be most strikingly illustrated by studying the action of the fibrinogen on the intravascular blood of the dog.

Exp. A.—40 c.c. of a freshly prepared slightly alkaline clear

Q

solution of tissue-fibrinogen were injected into a dog. (The tissue-fibrinogen was prepared as above indicated—twenty-six hours' watery thymus extraction, acetic acid precipitation, solution, &c.) After a short interval of about one to one and a half minute the respirations stopped completely, although there were one or two subsequent respirations. The carotid, on being opened, yielded a drop or so of blood, but no more, and it was flaccid and empty. A large volume of blood, free from admixture, was collected from the greatly distended heart and from the vena cava. This blood was entirely fluid. A careful examination of the animal, made immediately after death, showed that the right heart contained a small quantity of shrunken and whipped clot. The main pulmonary artery branches were entirely free from clot; the left heart was quite free from clot; the large veins of the forelimb and superior vena cava, the iliac veins, were completely devoid of any coagulum. But the whole portal venous system, commencing in the venous radicles close to the intestine and stomach, the splenic vein, the main trunk of the portal vein and all its branches in the liver, were completely thrombosed. The hepatic vein and its branches were quite free from clot; none of the arteries in the body were thrombosed.

The shed blood was centrifugalised; the plasma, left to itself, coagulated completely after an interval of twenty-four hours had elapsed, but until that time there was no appreciable clotting. It clotted rapidly on addition of some of the solution which had been injected; it also clotted readily when leucocytes from lymph-glands were injected.

The above experiment is typical of what occurs when a solution of tissue-fibrinogen is injected in quantity into a dog, and it is an experiment which it is extremely easy for anyone to repeat. The pronounced tendency to form clotting in the portal system is always very apparent. The degree to which clotting occurs elsewhere, especially in the right heart and pulmonary artery, varies in different animals and with varying conditions of diet.

The results obtained by a large number of experiments can

be summarised thus. If a solution of tissue-fibrinogen be injected into a dog in varying quantity the effects observed are : with very small quantities, no discoverable intravascular clotting occurs, but the blood, drawn off after the injection, clots very slowly, one to two hours intervening; with larger quantities, intravascular clotting takes place, being, as a general rule, chiefly confined to the portal venous system, the extent of clot being greater as more tissue-fibrinogen is injected. The shed blood will not clot. The more tissue-fibrinogen there has been injected the more complete is this prevention of the clotting, the interval between the drawing off and the clotting varying from two to thirty hours. In most cases the blood can be readily made to clot firmly by additions, such, for instance, as the ordinary fibrin ferment, and in the great majority of cases it clots firmly on standing; so that this non-coagulation of the shed blood is not due to the exhaustion of the fibrin-forming substratum of the plasma, a result which would be utterly improbable from the very limited amount of intravascular clotting, and from the fact that very marked slowing of the shed blood is observed when no intravascular clotting whatever has been produced, as when small quantities only have been injected.

This delayed clotting of the shed blood has always been prominently mentioned in my previous publications,[1] and it is of extreme importance. Nevertheless, it has been much overlooked. The non-clotting of the shed blood is the negative phase of the fibrinogen interaction.

I have already stated that the blood drawn off after injection clots of itself after a longer or shorter interval. From the considerable difference in the firmness of the clot so produced, I have no doubt that if the amount of fibrin were determined in each case, it would be found to be extremely variable, and in the main to vary inversely with the quantity of tissue-fibrinogen injected. But I have not done this, partly because it presents great technical difficulties, and partly because, from subsidiary reasons which I do not here adduce, the result would not be

[1] 'On Intravascular Clotting,' *Proc. Roy. Soc.* 1886. [Collected Papers, p. 135.]

wholly intelligible. I will now take an extreme case, which
illustrates completely the particular point I am anxious to
emphasise.

Exp. B.—The same solution of tissue-fibrinogen was used
as in Experiment A. The weight of the dog was rather
greater, 1½ kilo. more. It is to be noted that to produce a
given effect with tissue-fibrinogen solutions, the larger the
animal the greater the amount of substance which must be
injected.

50 c.c. of the solution were injected; the animal recovered,
although the pulse and respiration were greatly affected at the
time of the injection. After an interval of three to four hours, a
second injection of 50 c.c. of the fibrinogen solution was made.
The fibrinogen was the same as in the first injection, but the
solution was stronger. The injection had no obvious effect.
After some minutes the animal was bled from the carotid, the
blood flowing freely in a strong stream. The examination of
the animal showed very extensive clotting in the portal system ;
but with the exception of one or two small venous radicles, the
clotting was certainly not due to the last injection, but to the
first, since the thrombi were distinctly decolourised. There was
only a very minute trace of clotting in the heart, and the blood
in all the great vessels was quite fluid ; so that the total quantity of fibrin formed from this blood and from the tissue-fibrinogen—which, as I shall have to point out, with great probability
enters mainly into the substance of the clot—was limited to this
clot in the portal system. The shed blood, however, would not
clot at all. The addition of leucocytes had no effect on it, nor
had the addition of more tissue-fibrinogen.

The plasma of the blood showed no trace of clotting after
three days. The plasma of the blood contained none of the
fibrinogen injected. The fibrinogen injected is precipitated
by acetic acid and is not dissolved by great excess of the acid.
Thus some of the tissue-fibrinogen diluted twenty-five times
gave a permanent perfectly obvious precipitate, with acetic acid
of 50 per cent. added as 1 to 4. The plasma of the blood was
perfectly clear with this addition. The fibrinogen of the blood

is precipitated by acetic acid, but it is readily soluble in excess.

Such an experiment as the above was made for a special purpose. It cannot be always repeated with success, for the reason that the first injection frequently kills the animal; but I have succeeded in carrying it out sufficiently often to be sure it is not an exceptional instance. And, further, the injection of a very large quantity of tissue-fibrinogen always leads to the production of a shed blood entirely non-coagulable, either spontaneously or on addition of leucocytes or of tissue-fibrinogen. The only difficulty is this, that frequently the circulation is arrested before the required quantity is got in, and then only marked slowing of the shed blood is produced.

To return to the plasma in Experiment B.

I have already pointed out that it does not contain any of the injected fibrinogen. But it contains a large quantity of substance which, although it is certainly different from the fibrinogen of ordinary extravascular blood because it will not clot by ordinary means, yet retains the prominent characteristics of this substance. It must be also borne in mind how limited the actual clotting of the blood has been in this case, and that the injected fibrinogen has disappeared in the process.

The fibrinogen of blood-plasma is precipitated by rendering the plasma decidedly acid with dilute sulphuric acid (0·4 per cent.). The plasma of blood in Experiment B is also precipitated by this means, and gives a very large precipitate. The fibrinogen of extravascular blood is precipitated by saturating with sodium chloride, and is totally insoluble in saturated sodium chloride solution. The plasma of Blood B on saturation with sodium chloride gives a proteid precipitate which, when completely washed with saturated sodium chloride till disappearance of proteid reaction, the salt removed, and the residue dried and weighed, amounts to 0·93 per cent.; and since both the other known proteids in blood-plasma, paraglobulin and albumin, are soluble in saturated solution of sodium chloride, the 0·93 per cent. corresponds to that amount of fibrinogen in an altered form.

To summarise: What can be observed by the action of tissue-fibrinogen on blood?

If a certain quantity of tissue-fibrinogen interact with circulating blood a limited intravascular clotting will take place. Roughly, the amount of the intravascular clotting is proportional to the amount of fibrinogen injected. Simultaneously with this intravascular clotting another process takes place —the inhibition of the clotting of the shed blood. And it is quite certain that this inhibition is the more complete the more tissue-fibrinogen is injected. The injected tissue-fibrinogen, so far as can be ascertained by chemical and physiological tests, disappears wholly, and hence is probably the main substratum of the fibrin formed intra venas. The fibrinogen of the blood is profoundly altered. That it enters to a certain extent into the clot is probable, but a large quantity is certainly left.

This interaction (in which the injected substance disappears, in which the fibrinogen of the blood is altered, in which there is—I., clotting; II., inhibition to clotting, in which extent of thrombosis and diminished *coagulability* are quantitatively closely associated) I call ' fibrinogen interaction, with a positive and a negative phase.'

I regard this process as being of very great importance, because I have strong reasons for thinking that it is essentially analogous to the process of all zymotic disease.[1] In what I have been describing above we have to deal with a poison which, whilst it produces a definite local effect, exhausts itself in the process; and, whilst producing this effect and exhausting itself, is *pari passu* protecting the organism by altering the one chemical body on which it can act—' the exhaustion of the soil ' of the bacteriologist.

In these experiments, in which the fibrinogen is added to

---

[1] *Vide* ' Note on Protection in Anthrax,' *Proc. Roy. Soc.* 1887 ; and ' Ueber Schutzimpfung auf chemischem Wege,' *Du Bois' Archiv*, 1888, p. 529. [Collected Papers, pp. 321, 329.]

the whole blood, it cannot, of course, be said with certainty that the results which take place when the fibrinogen is injected are limited to the interaction of the fibrinogen injected with the fibrinogen of the blood. But I am decidedly of opinion that such is wholly and entirely the case. Intravascular clotting has, however, been observed by other investigators, and also the delay in the clotting of the shed blood. Thus, the various pupils of Alexander Schmidt have asserted that fibrin ferment of a special character, or blood rich in ferment, causes intravascular clotting, and also slows the coagulation of the shed blood. The latter result, and in some cases the former, they refer, in accordance with the general views they hold on coagulation, to the action of the white blood-corpuscles. This view of the matter I hold to be entirely erroneous, and shall discuss it later. But at present it is more convenient to consider the action of tissue-fibrinogen on extravascular blood, and thus complete our information as to the action of this body.

In the description already given of the action of tissue-fibrinogen on intravascular blood, it has been stated that the shed blood obtained after the injection of a medium quantity of tissue-fibrinogen—i.e. a quantity sufficient to produce a very pronounced effect (thrombosis) without being such as to kill the animal at once—clots very slowly. It clots, however, rapidly as soon as tissue-fibrinogen is added to it. The same is the case with the plasma, free from all form-elements, so that there is no question that in the extravascular interaction the presence of form-elements is not necessary. It is further to be noted that the more tissue-fibrinogen injected the less readily and completely does the shed blood clot on further addition of this substance, so that after very large doses or double doses it will not clot at all.

The most convenient plasma for studying the question is peptone-plasma. It is quite immaterial whether we consider this plasma as 'normal' or not. When solutions of peptone and tissue-fibrinogen come together an interaction takes place. Our object is to understand the nature of this particular

interaction.   Peptone-plasma is, like all extravascular forms
of plasma, not a constant fluid, but varies more or less in
different cases.   I have already alluded to a variety which I
designate as 'strong peptone-plasma,' meaning thereby the
plasma obtained when the animal has been very fully peptonised.   To a plasma of this nature tissue-fibrinogen solution
was added in the following experiment.

Exp. A.—To 20 c.c. plasma 10 c.c. of tissue-fibrinogen
solution.   The mixture becomes quite solid in three to four
minutes.   The clot is squeezed out; the serum on standing
does not clot any further.   On adding more tissue-fibrinogen
no further clotting occurs.   The serum is quite clear.   On
adding acetic acid in excess, a permanent precipitate is produced (serum 3 parts, acetic acid 30 per cent. 1 part).   The
original plasma, with this degree of acidity, is perfectly clear.

On adding this serum to a further portion of plasma clotting
occurs.   The serum from this clotting is perfectly clear with
strong acid.

In Experiment A clotting occurs; the tissue-fibrinogen is
still found in the serum, recognised by its insolubility in strong
acetic acid and by its causing more plasma to clot.   .

Exp. B.—The tissue-fibrinogen solution is now diluted four
times, and to 30 c.c. of the plasma 7 c.c. of the fibrinogen solution is added.   Coagulation takes place, the whole becoming solid.
The serum of this remains uncoagulated.

When tissue-fibrinogen is added, it again clots completely.
The serum from this remains uncoagulated.

It remains clear with strong acetic acid; it does not cause
clotting when added to new plasma.   It gives a scarcely perceptible clot when new tissue-fibrinogen is added.

To 10 c.c. of the plasma 1 c.c. of the fibrinogen solution is
added.   A very scanty, imperfect trace of clotting occurs.

The plasma, before adding the tissue-fibrinogen, coagulates
slowly but firmly on passing a current of CO_2 and on diluting.
After the addition of the tissue-fibrinogen it would not clot with
these or other means.

The plasma gives a precipitate on acidifying with dilute

sulphuric acid; and it can be shown (though it was not done in this experiment) that this precipitate is the substance which interacts with the tissue-fibrinogen. After the plasma has been clotted by the addition of tissue-fibrinogen in proportion just sufficient to leave no excess of the latter, it still gives a large precipitate with dilute acid.

The process which takes place between the plasma and the fluid is not of the character of a fermentative process, and for the following reasons :—

1. The process is obviously more or less quantitative.

2. The tissue-fibrinogen disappears in the process.

3. The power of the plasma to cause the disappearance of tissue-fibrinogen is strictly limited. So soon as any excess is added it remains in the serum quite unaltered.

4. The plasma itself is profoundly altered in the change. It contains a non-coagulable fibrinogen.

I purposely avoid discussing in great detail the behaviour of extravascular plasma and fibrinogen solutions till a later section of this report, since it involves the discussion of many points which I reserve for the chemical section.

If we compare the action of tissue-fibrinogen on the circulating blood of the dog and its action on the extravascular peptone-plasma of the same animal, from which all form-elements have been removed, we find that there is at all events a great resemblance. It is, therefore, I think, quite unnecessary to assume that the intravascular coagulation and its accompanying inhibitory phase can only be explained by introducing white blood-corpuscles into the process.

As I have already stated, other observers have noted that after intravascular clotting the shed blood has clotted slowly. Among the earliest of these observations on intravascular clotting are those of Köhler.[1] This observer uses what he terms ' ferment-rich blood.' To obtain this the blood is allowed to clot, and the clot, whilst still warm, is squeezed out through linen. The blood so obtained produces sometimes marked intravascular clotting. This result is attributed to the fibrin ferment. If this method be

---

[1] Armin Köhler, ' Ueber Thrombose,' &c., Inaug. Dissert., Dorpat, 1877.

repeated, it is found that such a conclusion is in no way justified.
Blood so treated, if centrifugalised, invariably gives a serum
intensely coloured with hæmoglobin, and containing stromata.
Now I have shown that the injection of stromata of the mammalian blood-corpuscles causes intravascular clotting,[1] and this
result has been confirmed by Krüger as regards the stromata of
birds' blood.   It is not contended by anyone that the stromata
are identical with fibrin ferment.

This so-called ' ferment-rich blood ' contains, in addition to
the stromata, an unusually large amount of serum-fibrinogen.
Possibly it contains a large amount of fibrin ferment ; but, since
it has been repeatedly observed that the injection of solutions of
ferment, which are very efficacious on the test fluid, are without
influence on intravascular blood, it would be obviously un-
justifiable for Köhler to apply his results to the fibrin ferment.

The only attempt which has been made, so far as I know, to
explain or to analyse this slow clotting of the shed blood has
been to suppose that the injection has destroyed the white blood-
corpuscles, and that the lack of these bodies in the shed blood
is the cause of its not coagulating.   Such an explanation can
hardly be satisfactorily applied to the phenomena which I have
described with regard to the action of tissue-fibrinogen on the
dog.   It is true that the shed blood, after moderate injections,
clots readily on the addition of leucocytes from lymph-glands.
After large injections, however, the addition of leucocytes or
tissue-fibrinogens is wholly without influence, in spite of the fact
that the intravascular clotting is of so limited a nature that a
total exhaustion of the fibrinogen of the blood is impossible.
The simple destruction and absence of white corpuscles as a
cause of the non-clotting of the shed blood is excluded by these
experiments, and any further way in which they could affect the
result is a matter at present of pure hypothesis.

As a matter of fact, white blood-corpuscles are invariably
present in the blood after the injection of tissue-fibrinogen.
Their exact enumeration and comparison with those in blood
previous to the injection is very difficult.

[1] *Practitioner*, March, 1886.   [Collected Papers, p. 167.]

In all cases of slow-clotting blood, such as is obtained after the injection of tissue-fibrinogen, precipitations of fibrinogen (or, at any rate, of granular matter, which I take to be fibrinogen) occur. These may be microscopic in quantity or considerable enough to form an obvious coagulation. Thereby an extremely irregular distribution of the white corpuscles occurs. I have, for instance, in peptone-blood—the non-coagulability of which has been explained by Samson-Himmelstjerna,[1] a pupil of Alexander Schmidt, as being due to its containing practically no white corpuscles—repeatedly observed six or eight fields and found no white blood-corpuscles, in a ninth a mass of at least eighty, and I have made similar observations on the shed blood after tissue-fibrinogen injections. Under these circumstances the enumeration of white blood-corpuscles is hardly a reliable method of ascertaining whether a destruction has taken place. It is at present the only evidence afforded. In the case of serum-fibrinogen in its behaviour on intra- and extravascular plasma, in the case of the action of leucocytes, in the case of tissue-fibrinogen in its action on intravascular blood and extravascular plasma, I have endeavoured to illustrate what I call ' fibrinogen interaction,' a process with two opposite phases, and not a fermentative process.

Pseudo-Fermentative Action and Progressive Coagulation.

Under this heading I shall describe a process which is of very great importance in the phenomena of coagulation. The exact analysis of the matter is difficult, and I am not at present able to treat it in detail.

The quantitative relationship, which can be so clearly shown in dealing with the interaction of strong peptone-plasma and tissue-fibrinogen, does not prevail in all kinds of plasma. If the peptonisation is less complete, the addition of tissue-fibrinogen causes a coagulation which is quite out of proportion to the amount of tissue-fibrinogen added, and instead of the process ending with the first coagulation, it is continuous—*i.e.*

---

[1] Samson-Himmelstjerna, Inaug. Dissert., Dorpat, 1882.

the serum from the first clot after a short interval clots again, and, it may be, again. Such a result might easily lead to the conclusion that the tissue-fibrinogen really acted as a ferment, for the fact of the substance being used up, which is always the case in these progressive coagulations, is not so clearly shown as when larger quantities of the solution are added. The complication of this process is, however, fully brought out by the fact that prolonged boiling of the tissue-fibrinogen in no way impairs its power. The fact that boiled solutions of tissue-fibrinogen cause clotting in extravascular blood has been long pointed out by me, and not only do the solutions do this, but the coagulated tissue-fibrinogen is extremely effectual [1] in all varieties of peptone-plasma. The action of these boiled solutions of tissue-fibrinogen will be more fully discussed in the chemical section of this report. It does not act in the same way as the unboiled; but, as far as I can at present state, the boiled tissue-fibrinogen, like the unboiled, becomes a part of the fibrin formed. This is, however, not the case with the coagulated tissue-fibrinogen, which, all the same, causes clotting.

It would, therefore, be erroneous to rank this progressive coagulation under the head of a process due to the action of an enzyme.

In view of what I have stated, there is the possibility that a true fibrin ferment does not exist. So far as my own knowledge goes, I am of opinion that there is a real enzyme connected with fibrin formation—the fibrin ferment of Alexander Schmidt. I have not yet, however, completed my observations in this direction.

In the account given of fibrinogen interaction I have discussed a chemical process, employing terms more often applied to the physical side of physiology. This is because I do not attempt, in this part of the report, to enter into chemical details. It must, however, be understood that, although the problem is no doubt a chemical one, it is of so complex a nature that chemical means alone are quite inadequate to investigate it, and

---

[1] 'Ueber Schutzimpfung auf chemischem Wege,' *Du Bois' Archiv*, 1888. [Collected Papers, p. 329.]

that only here and there are we at all able to attack it chemically. It is highly probable that the action of the vagus on the heart is associated with chemical alterations of the heart's muscle. The intimate nature of these chemical processes, we may safely assume, will not be known for some years.

Another point, on which confusion as to the action of ferments and fibrinogen interaction might arise, is in the fact that certain ferments—e.g. diastase—are known to produce a dissociation which might be taken as analogous to the positive and negative phase of fibrinogen interaction. From starch we get sugar and dextrin as the result of the action of diastase, but we know that neither the sugar nor the dextrin are formed out of the diastase. In the fibrinogen interaction, a strictly quantitative process, we get fibrin formed and another fibrinogen. Both bodies are different from the original fibrinogens, but both the fibrin and the altered fibrinogen are derived from the original fibrinogens.

The Fibrin Ferment and the Normal Coagulation of the Blood

In all the processes which I have been discussing, the fibrin ferment (as defined by me) plays no part.

The serum-fibrinogen is without action on the ferment test fluid. The isolated leucocytes are totally without action on the test fluid. The tissue-fibrinogen is totally without action. The intravenous injection of ferment (*vide infra*) is without influence. The intravenous injection of all the above bodies is of great influence. Strong peptone-plasma is not affected by ferment, but is affected by all the above bodies.

The fibrin ferment is a substance which is found in blood-serum, the product of coagulation; in the blood as it leaves the vessels it is absent. It is, therefore, and always has been, an ungrounded assumption to suppose that the ferment is the cause of coagulation.

The modern investigations on this subject, even those by the most ardent supporters of the ferment doctrine, have not tended to diminish the difficulties of this assumption. My firs experiments—on the action of leucocytes on peptone-blood [1]—showed very clearly that there were great difficulties in the way of supposing that the action of leucocytes consists in their discharging into the plasma the lacking constituents which were found in serum of blood, since peptone-blood, readily affected by leucocytes, is unaltered by serum.

Rauschenbach,[2] who confirmed and extended my results, made the observation that the cells contained no fibrin ferment, and the filtered plasma was equally deficient in this substance. But on bringing the cells and the plasma together coagulation takes place, and the serum from this coagulation contains much ferment. This is, in fact, precisely the state of things in the normal coagulation of the blood. The blood, directly it leaves the vessels, white blood-corpuscles and all, is free from fibrin ferment. Clotting ensues, and there is plenty of this substance. Plainly, therefore, there is some chemical process in coagulation apart from the action of the ferment upon fibrinogen. This chemical process was previously placed by Schmidt in the protoplasm of the leucocytes, and he makes various suppositions as to the white blood-corpuscles breaking up in different ways, now yielding, and now not yielding, ferment. All this was pure hypothesis, and was never clearly defined. Now that I have introduced the method of working with plasma separately, and with cells separately, this idea is wholly given up, and from now onwards it is recognised by all the Dorpat workers that both plasma and cells are engaged in the interaction which produces the ferment. It is always a question of ferment sources, from which the plasma 'splits off' the ferment.[3]

Three statements have been made by the Dorpat school as to the nature of this process :—

1st. That fibrinogen is not concerned in the process (Rauschenbach).

2nd. That possibly the minute traces of ferment contained

---

[1] *Du Bois' Archiv*, 1881. [Collected Papers, p. 69.]

[2] *Blutplasma und Protoplasma*, Dorpat, 1883.

[3] Nauck, Dissertation, Dorpat, 1882.

in plasma split off greater quantities of ferment from the form-elements (Nauck, 1886).

3rd. That possibly an unknown body in the plasma is concerned in this production of ferment (Nauck).

This was the state of affairs when my 'Croonian Lecture Abstract,' and 'Uebersicht einer Theorie der Gerinnung' in Ludwig's 'Festschrift' were written.

But in the meanwhile I had drawn attention to another aspect of the question.

In my paper on the 'Origin of the Fibrin-Ferment'[1] I pointed out that in spontaneously coagulable plasma, perfectly free from form-elements, the *formation* of fibrin ferment goes on. Ten per cent. sodium chloride plasma and peptone-plasma are free from fibrin ferment. They can be made to coagulate by simple means, CO_2 and dilution. Coagulation occurs, and the serum contains fibrin ferment. In testing for the ferment I use the method of Alexander Schmidt. We see, therefore, that this process, antecedent to the formation of the fibrin ferment, is not a process in which form-elements of any sort are necessarily concerned, since its formation goes on in plasma perfectly free from form-elements. I cannot state whether, as a general rule, the amount of ferment which is developed in plasma is less than that in normal serum, because clotting may occur without any detectable amount of the fibrin ferment being formed at all, and there are probably special conditions which favour the production of a large or small quantity of fibrin ferment. But in numerous cases the amount of ferment formed as the result of the coagulation, or of connection with plasma free from form-elements, is quite equal to that of the normal serum of the same blood.

To take now the case of leucocytes. I am quite of the opinion of Rauschenbach that they contain no fibrin ferment, since they have no action on dilute magnesium sulphate plasma, the test fluid for ferment; and I am equally certain that when they cause coagulation in peptone-plasma free from ferment, ferment is found after the clotting. The same observation has

[1] *Proc. Roy. Soc.* 1886. [Collected Papers, p. 119.]

been made by Rauschenbach on cooled plasma. I am directly opposed to the conclusion of Rauschenbach that the fibrinogen is not the substance in the plasma which interacts with the cells. Rauschenbach supports this view by the fact that, though the filtered cooled plasma of the horse clots readily and forms ferment with leucocytes, the pericardial fluid of the same animal does not clot with leucocytes, although it readily does with ferment; hence he concludes that it cannot be the fibrinogen of the plasma which affects the interaction with the cells, since the pericardial fluid also contains the fibrinogen and yet does not interact.

I have pointed out, however, that the !fibrinogen of the blood cannot be looked upon as the more or less constant body which Schmidt and Hammarsten have supposed it to be. I have pointed out that the precipitated fibrinogen of the plasma interacts with the cells, but that in the process of precipitation it undergoes marked changes, which are still more increased by re-solution and re-precipitation, until, if the process be continued, a fibrinogen is obtained which has not the slightest interaction with leucocytes. Such a fibrinogen coagulates, however, very readily with fibrin ferment. In fact, such a fibrinogen resembles closely the fibrinogen of transudation fluids. That fibrinogen of the blood is most profoundly altered by precipitation can be readily seen by anyone. For, if any one will precipitate the salt-plasma of the dog in three cases, by means of saturation or semi-saturation with common salt, he will find that in two out of the three cases the precipitate is no longer soluble in dilute salt solutions, and consists largely of a substance indistinguishable from fibrin.

In the case of the horse, however, this change is more gradual, so that the different characters of the precipitate, both as regards its solubility and its behaviour to clotting agents, may be easily followed.

That the fibrinogen of the plasma is the changeable substance I described it is now practically admitted by the Dorpat school. At any rate, they have given up the term ' fibrinogen ' altogether, and speak now of ' metaglobulin.'

The Dorpat school then allow that there is some chemical process antecedent to coagulation—*i.e.* the formation of ferment—and that in this process the plasma is concerned. Hence there is at present no upholder of the strict fermentative view.

There is, however, another point to be considered. It is admitted by everyone that fibrin ferment has no influence on the circulating blood, at any rate, of certain animals, such as the dog. Now I have shown that the fibrin ferment is equally inactive on the extravascular plasma of the same animal, at all events under certain conditions. We have under the ferment doctrine of coagulation, therefore, not only to consider some chemical process giving rise to the ferment, but the possibility at least of some further chemical change, by which the blood is made suitable to react with ferment. The ferment doctrine of coagulation is, therefore, far from affording a satisfactory explanation of the phenomena of normal coagulation.

In the previous part of this section I have described the phenomena under the head of ' fibrinogen interaction.' This process is a different one altogether from a fermentative process. Nevertheless it produces fibrin, and it produces fibrin under conditions in which the fibrin ferment will not produce it, and in many cases, moreover, particularly in extravascular plasma, fibrin ferment is one of the products of fibrinogen interaction. I have, therefore, adopted the view that normal coagulation is a fibrinogen interaction, and that the fibrin ferment is one of the by-products of this interaction.

If, for instance, coagulation be set up in peptone-plasma by tissue-fibrinogen or serum-fibrinogen, not only does fibrin make its appearance, but also fibrin ferment.

The same is the case when the plasma is made to coagulate by a current of CO_2 ; here there occurs an interaction between two fibrinogens present in the plasma, and both fibrin and ferment are formed. But peptone-plasma is not affected by ferment, and the process of coagulation which occurs when fibrinogen is added to peptone-plasma does not in the least resemble a ferment action. Nor is the coagulation which occurs

R

in peptone-plasma as the result of CO_2 always accompanied by the formation of ferment. It may be entirely absent from first to last, and the quantity of ferment formed can at will be made to vary within very wide limits without the rate of coagulation being perceptibly altered. I think, therefore, that it must be admitted that there are forms of coagulation in which the ferment is a wholly unimportant by-product of a chemical interaction—fibrinogen interaction—in which fibrin is produced.

Let us suppose for a moment that white blood-corpuscles are concerned in coagulation. From our acquaintance with the action of leucocytes on plasma and blood, it is certain that these cells, considered as matter, act not as ferments, but as fibrinogen. Leucocytes produce a coagulation in extravascular plasma. They do this in extravascular plasma utterly unaffected by fibrin ferment. In intravascular plasma they produce no coagulation. If injected in sufficient quantity they stop it. We understand a ferment which produces a given change; we do not know of any which has also the power of producing the exact opposite.

For special reasons, which I have already adduced, I think we must look upon the white blood-corpuscles as differing from leucocytes of lymph-glands, because, since they have already entered the blood-current, they have already undergone the negative fibrinogen interaction, and are therefore, I think, excluded in extravascular clotting. If the white blood-corpuscles have any influence at all on coagulation, it is that on entering the blood-current they tend to keep the blood fluid, just as the injection of leucocytes keeps the blood fluid. In all probability (I have reasons for so saying) the normal fluidity of the blood is due to the negative fibrinogen interaction of the blood and the vascular wall (endothelium).

The impetus to coagulation cannot, therefore, start from a fibrinogen interaction between white corpuscles and plasma. The idea that the white corpuscles explode or break up, and simply liberate ferment, is an hypothesis which has been abandoned by every intelligent worker on coagulation. The processes of coagulation must therefore take place in the plasma, and it is, accordingly, of importance to ascertain whether, in the plasma, the conditions are present for a fibrinogen interaction, such as occurs when tissue-fibrinogen is added to peptone-plasma or to the plasma of circulating blood. I have endeavoured to show that this is the case by the study of the coagulation of peptone-plasma and salt-plasma. The fibrinogen interaction explains, at any rate, a great deal of coagulation, whereas the production of fibrin by the action of ferment on fibrinogen cannot be more than the last of a series of important chemical processes, which are not explained at all by the ferment doctrine.

I shall not repeat the evidence I have adduced in separate papers as regards the fibrinogen interaction of the plasma. I will, however, lay stress on the following points, which can only be dealt with in a fragmentary manner in this section of the report, their full discussion being more fitly reserved for the section on intravascular clotting.

Peptone-plasma, as usually obtained, is spontaneously coagulable on the passage through it for a few minutes of a current of CO_2; this power is lost or greatly diminished by cooling the plasma and removing the substance produced by the cooling. Very often a plasma is obtained which clots very slowly and very badly with CO_2. Such a plasma separates very little substance in the cold. The conditions which favour this or that plasma are very complicated. They depend on the dose of peptone; on the condition of nutrition of the animal; on the time of the last meal. The substance which separates on cooling can, with right, be spoken of as fibrinogen, since it is converted with great readiness into fibrin. To distinguish it, I have called it ' A-fibrinogen.'

No matter how long the cooling may be, the plasma still contains the bulk of the coagulable matter (fibrin-yielding), and the remainder of the fibrinogen which does not separate from the plasma by cooling I distinguish as ' B-fibrinogen.'

Now every peptone-plasma coagulates readily with tissue-fibrinogen, and the plasma-fibrinogen, as such, disappears, as does also the tissue-fibrinogen.

R 2

From peptone-plasma, by saturating with neutral salt, a precipitate can be obtained. As I have mentioned above, the character of this precipitate varies greatly, and I do not know by what means it could be assured that the precipitate should always have a definite character.

If this precipitate—which is fibrinogen, since it is converted easily into fibrin—be added to peptone-plasma, it *may* cause clotting, or it *may* prevent plasma which, before the addition, was easily coagulated by CO_2, from clotting by this means. That is to say, the alteration the fibrinogen undergoes in precipitation is such that, when added to the same plasma, it acts as if it were a new fibrinogen, producing now a negative, now a positive predominating phase.

The action of carbonic acid on peptone-plasma is, at any rate, partly due to the fact that it tends to make some of the fibrinogen of the plasma less soluble. The fibrinogen of the plasma is, as I have stated, precipitated by acid. It is also precipitated by cooling; the lower the temperature the greater the precipitate. CO_2 by itself causes no immediate precipitate in peptone-plasma, but it greatly favours the precipitation of the body A-fibrinogen in the cold.

I must here mention that the precipitation of the A-fibrinogen by cold is invariably attended with a change, just as the precipitation by salt invariably alters the fibrinogen of the plasma. This change is shown by its altered solubility and by its being converted into a fibrinous body, in many cases absolutely undistinguishable from true fibrin.[1] If, now, the coagulation of the peptone-plasma is closely connected with the precipitation and alteration of A-fibrinogen, warmth ought to be disadvantageous to the clotting of the plasma by CO_2, since warmth tends to keep the A-fibrinogen in solution.

If peptone-plasma be warmed to 37°, and maintained at this temperature during the passage of the CO_2, and kept at this temperature subsequently, it will not clot; although, if it be treated

---

[1] Freshly precipitated, it is easily soluble on warming, but rapidly loses this power. Easily soluble in 4 per cent. NaCl, it rapidly loses this power. Easily soluble in very dilute alkalies, and it quickly loses this power.

with CO_2 at the ordinary temperature of the room, it clots readily. I have repeatedly made such experiments. I now describe two others.

(1) Peptone-plasma of the strong variety was treated with CO_2 at the temperature of the room. Repeated applications of the CO_2 only produced a mere shred of fibrin. The plasma was put for a few minutes into the ice-chest. It was just becoming slightly turbid when removed. On treatment with a current of CO_2 it clotted through and through.

(2) Peptone-plasma, which was at first readily coagulable with CO_2, had been cooled for twenty-four hours and the precipitate removed. It was then again allowed to stand twenty-four hours in the ice-chest. A further slight precipitate had formed and partially subsided, so that the upper half of the tube was quite clear and the lower half very slightly turbid. The upper half was removed by a pipette. It would not clot with repeated administrations of CO_2. The lower half clotted with CO_2, the clotting commencing in about fifteen minutes, and going on in crops.

The points which I am endeavouring to elucidate are as follows : Peptone-plasma remains for long periods unclotted. It undoubtedly clots when tissue-fibrinogen, or serum-fibrinogen, or fibrinogen prepared from peptone-plasma is added; it also undoubtedly clots as the result of a current of CO_2. The CO_2 coagulation is also apparently a clotting due to fibrinogen interaction, for the removal of a fibrinogen stops it (A-fibrinogen); and warmth, a condition which hinders precipitation and consequent alteration of fibrinogen, also stops it. It must be distinctly understood that the preventive influence of warmth is only exerted when it is applied previous to the CO_2 action. If clotting, or even precipitation, has once occurred, warmth exerts a decidedly accelerating effect.

A very frequent phenomenon of the coagulation of peptone-plasma is the fact that it coagulates with CO_2 in crops. It first clots through and through. On squeezing out this clot, the serum obtained clots again. Previous to the clotting of this serum it becomes turbid and a precipitate is formed. The same may occur a third and fourth time.

The precipitates which occur in peptone-plasma as the result of cold and as the result of CO_2, and crop-like coagulation, are distinguished by the remarkable disc-like character of the granules. The primary cold precipitate I have frequently described. The precipitate I mention as occurring after CO_2 is also in the form of discs, but they are, as a rule, smaller and more sharply defined than the original cold precipitate. They also behave somewhat differently towards reagents. A similar precipitate is often observed before the beginning of the first coagulation with CO_2.

It is impossible that the fibrin ferment should play an important part in the coagulation of peptone-plasma. It is true, as I pointed out, that, after the coagulation of peptone-plasma by carbonic acid, ferment is found in the plasma. An appreciable quantity is only found in those cases in which there is a large quantity of the cold body.

I have investigated with great care the amount of ferment present in peptone-serum after coagulation, and in a large number of cases have found that coagulation has taken place with tolerable rapidity, but that there has been no detectable quantity of ferment formed.

I give the following example :—

Peptone-plasma quite clear first day, treated with CO_2. Coagulation began in ten minutes; went on in crops. Fibrin and serum under alcohol for one month. No. 1.

Same plasma, second day, imperfect cooling, not quite clear. Coagulation began rather more quickly. Alcohol for one month. No. 2.

Normal serum of dog. Alcohol for one month. No. 3.

The alcohol coagulum, after drying (*sec. art.*) in the three cases, was tested for ferment with dilute $MgSO_4$ plasma of the dog. Five grms. of each used, and extracted for thirty minutes with water (ten c.c. water). Five c.c. of the ferment solution added to an equal volume of diluted plasma. In No. 3, with the normal serum clotting took place in three minutes. In both the first, no clotting in twenty-four hours.

One grm. of the dried powder of each of the first caused no clotting whatever in twenty c.c. diluted $MgSO_4$ plasma.

One grm. of the dried powder of the normal serum caused very rapid coagulation round the powder, which rapidly extended through the fluid.

Such experiments I have frequently repeated. In cases where there is an abundance of the cold body present in the plasma, the ferment formed may be quite equal in quantity to that of normal serum. As, however, its appearance is inconstant, it appears reasonable in the case of peptone-plasma to regard the fibrin ferment as an occasional result of coagulation, for it is not present before coagulation in any case. The plasma may be greatly altered by artificial means—e.g. dilution—and thus made susceptible to the action of the ferment. But there is no ferment formed until actual coagulation occurs, and then it is an inconstant phenomenon.

It can, of course, be objected that peptone-plasma is an exceptional variety. Since it presents many points in common with the plasma in the vessels on the one hand, and with cooled plasma on the other, it cannot be so utterly abnormal as some of my critics have suggested. The same objection can with equal validity be applied to every form of artificial plasma. I will, for instance, shortly refer to the plasma which can be obtained by suspending the jugular vein of the horse, removed from the recently killed animal. It is, of course, utterly unadapted for exact experiments, since a complete removal of the form-elements is only effected with great slowness. Frédéricq, who has investigated this plasma with great care, says that even after five days suspension there are still leucocytes in the upper layers of the plasma. According to this author, almost without exception a small clot is formed at the junction of the red corpuscle column with the plasma. (The Blutplättchen controversy had not, at the time this author wrote, arisen, and hence he regards every granular collection as an indication of broken-up leucocytes.) Now, if clotting, however limited in extent, has taken place in plasma, it is no longer normal. I have related above that the injection of a small quantity of serum-fibrinogen profoundly alters the whole blood of the animal, and serum-fibrinogen is a by-product of ordinary coagulation. Further, this plasma (of the horse vein) constantly, according to Frédéricq, contains fibrin ferment in small quantity. Fibrin ferment is not present in the normal circulating blood of the horse, and hence, in this plasma, chemical processes tending towards coagulation must have already taken place. This is quite in accordance with the existence of the small clot alluded to.[1]

It is, therefore, very difficult to assert with confidence what is the normal process of events in the coagulation of shed blood. All experimental methods of investigating the clotting of the blood are open to objection. It can be said, in such and such a plasma coagulation takes place in such and such a way; but to say, therefore the normal blood clotting takes place in this way is not a correct conclusion. The only real connecting link between the experimental methods and the normal process which we have at the present time is the microscopical examination of intravascular clotting and of shed blood. This method, utterly inadequate and misleading by itself, becomes of the greatest importance when controlled by experimental evidence. Modern microscopical investigation has established with fair certainty that an appearance of discs in the plasma is a constant accompaniment of coagulation. It is disputed whether these discs are previously formed in the normal blood, and it cannot ever be ascertained, since the blood cannot be investigated microscopically until it is in an abnormal condition—*i.e.* commencing stasis. It is also disputed whether these discs are a form-element or a precipitate. It is, however, certain that they are present in shed blood, that they disappear in the clotting process, and that all slow intravascular clotting exhibits, as one of its main features, a large appearance of these discs. Further, Löwit[2] has shown that slight injury to the vascular wall causes a sudden large appearance of discs, which must in this case be a precipitate.

In peptone-blood the visible appearance of fibrinogen discs (A-fibrinogen) is, if not a necessary condition of coagulation, at any rate a most favouring condition (*vide* influences of

Frédéricq, 'Sur la Constitution du Plasme sanguin.' Gand. *Archiv für exp. Path.* 1887.

temperature); and the nature of the coagulation of peptone-blood is that of fibrinogen interaction.

In a previous part of this report I have stated what my conclusions are as regards the Blutplättchen, and especially that they probably include different things. In describing the behaviour of peptone-plasma I have spoken of the substance A-fibrinogen. The microscopical appearance of this substance is such that it cannot be distinguished from the Blutplättchen which occur in peptone-blood. The substance A-fibrinogen can be collected, chemically examined, and its action on coagulation accurately estimated.[1] I have further spoken of precipitates occurring in peptone-plasma in the process of clotting. These precipitates are often in the form of regular disc-like granules. I mention this here, but must leave the nature of these discs and their connection with coagulation to a subsequent section.

[1] I have frequently described this ; vide 'Ueber einen neuen Stoff,' *Du Bois' Archiv*, 1883 ; ' Uebersicht einer Theorie,' Ludwig's *Festschrift*, Leipzig, 1887 ; ' Beiträge zur Frage der Gerinnung,' *Du Bois' Archiv*, 1888. [Collected Papers, pp. 124, 186, 253.]

CLASS VII

FURTHER PAPERS, INCLUDING
THE UNFINISHED SECOND REPORT TO THE
GROCERS' COMPANY, DEALING WITH THE QUESTION
OF THE INTERACTION OF FIBRINOGENS
AS THE CAUSATIVE FACTOR
IN FIBRIN FORMATION

CONTRIBUTIONS TO THE COAGULATION QUESTION[1]

I. On the Relationships between Fibrinogen and Fibrin

IN my paper ' Uebersicht einer Theorie der Blutgerinnung,'[2] I endeavoured to show that the precursors of fibrin are not pure proteids, but substances which contain proteid and lecithin. They are known by the name of ' fibrinogens,' and are found not only in the blood, but may be obtained from almost every animal tissue, such as the thymus, testis, brain, liver, the stroma of the red blood-corpuscles, and so forth. These proteid-lecithin compounds I have designated ' tissue-fibrinogen.' On coming in contact with blood-plasma they are converted into fibrin both within the living circulation and outside the vessels.

The fibrinogens of different origin vary in their behaviour towards a number of reagents ; on the other hand, they all possess in common numerous properties characteristic of the class. In the course of this paper I shall mention a number of properties common to all, and shall more especially dwell upon their relation to fibrin. The fibrinogens referred to are those obtained from blood-plasma, testis, thymus, as well as that from the stromata of the red blood-corpuscles. Fresh fibrinogens form apparently clear solutions in water, in dilute alkalies, and in dilute salt solutions. The question whether they form true solutions must for the present remain in abeyance. If a 4 per cent. sodium chloride plasma be prepared from blood taken from a well-fed animal, and be filtered through a clay cell, no fibrinogen passes through. If, however, the experiment be made with the blood of a fasting animal, fibrinogen passes

[1] [Translated from *Du Bois' Archiv*, 1888, p. 174.]
[2] Festschrift für C. Ludwig, Leipzig, 1887. [Collected Papers, p. 186.]

through. Similar observations have been made on the casein of milk.

All fibrinogens are very susceptible to the action of the various methods of precipitation. They cannot be thrown down without altering their properties, and especially their solubility. The stromata of the red blood-corpuscles are, for instance, soluble in water, or at any rate they swell up so as to cause an apparent solution; if precipitated by dilute sulphuric acid they lose their solubility (or power of swelling up). The fibrinogens from testis and thymus are obtained by finely chopping these organs and extracting with distilled water; they give clear solutions which may be filtered. When precipitated with acetic acid they become insoluble in pure water, and a little alkali or chloride of sodium must be added to bring them into solution. The fibrinogens of blood-plasma behave in a similar manner. The paraglobulin may be precipitated by diluting the peptone-plasma with ten times its bulk of water and by passing a stream of carbonic acid through it. After this has been done, the plasma may be diluted to any amount, without any fibrinogen being thrown down. Coagulation sets in only very gradually.

The fibrinogens may all be precipitated with acids, but to effect this the reaction must be markedly acid. If dilute mineral acids, and especially sulphuric acid, be employed, the precipitate redissolves in excess of the acid, and the more readily the shorter the time it has stood. After standing some time the solution is incomplete, and the fluid turbid.

If pepsin be added to solutions of fibrinogen which have been rendered clear by an excess of dilute mineral acids, and the mixture be allowed to stand for a few hours at 37° C., a voluminous precipitate is thrown down, which does not redissolve, however long digestion may be allowed to go on. So long as the precipitate remains fresh it is readily soluble in dilute alkalies, but not in dilute acids. In concentrated nitric acid the solution has a yellow or a greenish-yellow colour; on warming, and on adding ammonia, the xanthoproteic reaction is obtained. The precipitate gives a strongly acid ash on combustion. If a little soda and saltpetre be added before ignition, the ash is found to contain a large amount of phosphoric acid. The greater part, if not all, of the phosphorus arises from the lecithin of the precipitate produced by digestion. A comparatively large amount of lecithin can be extracted from the precipitate by means of alcohol, and this may be repeated until so little lecithin is left that mere traces of phosphorus remain in the ash. Iron is likewise invariably found in the ash. It cannot be extracted from the precipitate by means of alcohol and hydrochloric acid before ignition.

The precipitate arising from the digestion of fibrinogens resembles the bodies which Miescher [1] and Bunge [2] obtained by digesting yolk of egg, and termed ' nuclein.' When the nuclein contained iron, Bunge gave it the name ' hæmatogen.'

Under appropriate conditions the fibrinogens coagulate and form fibrin which likewise contains a considerable quantity of lecithin. When allowed to digest with acid and pepsin, the fibrin does not give rise to any precipitate, but dissolves, forming a clear liquid, which remains so, however long digestion may be allowed to proceed. In the coagulation of the fibrinogens a change must occur in the relation of the lecithin to the proteids, so that the compound can no longer be broken up by the peptic juice. It must be taken into consideration that the fibrinogens may leave a residue on conversion into fibrin. Under certain circumstances other proteids may arise as by-products, these being likewise soluble in the digestive juice.

The statement that fibrin may be dissolved by pepsin to a clear solution refers to the substance produced from isolated fibrinogens. If ordinary fibrin which has been obtained by whipping the blood be subjected to digestion, a considerable residue invariably remains. This is supposed by Hammarsten to be the products of the white blood-corpuscles entangled in the clot. This may be partially correct ; but the matter is capable of another explanation. As I have already shown on

[1] *Medicinisch-chemische Untersuchungen*, edited by Hoppe-Seyler, 1871, Heft 4, pp. 441 and 502 ; also Kossel, *Zeitschr. f. physiolog. Chem.* vol. iii. vii.

[2] *Zeitschr. f. physiolog. Chem.* 1884, vol. ix. p. 49.

several occasions, the appearance of fibrin is preceded by the separating out of discs which adhere readily to each other, thus forming masses that are but little affected when coagulation takes place. They are thus left as masses of fibrinogen which have been scarcely altered at all ; they are imbedded in the fibrin, and give rise to the above-mentioned precipitate on digestion.

I will now describe a few experiments which will serve to illustrate my remarks.

1. Horse's blood is received into magnesium sulphate solution and centrifugalised. The fibrinogen is then precipitated from the plasma by half-saturation with sodium chloride; the precipitate is filtered off, well pressed between filter-papers, and divided into two portions. Portion A is mixed with 0·2 per cent. HCl and pepsin. A considerable precipitate remains undissolved after three days' digestion. Portion B is dissolved in very slightly alkaline water and some horse's serum added. The fluid coagulates in about half an hour, and a solid clot is soon formed. This is cut into pieces and set to digest, together with the fluid which has been pressed out. At the end of twenty-four hours it has all dissolved, forming a liquid as clear as water, which does not become clouded even after the lapse of several days.

2. Peptone-plasma, which has been centrifugalised until it has become perfectly clear, is mixed with dilute hydrochloric acid (about 0·3 per cent.). The precipitate that at first occurs disappears as more acid is added ; some pepsin is then added to the clear solution. After standing for twenty-four hours in the warm chamber the mixture becomes white and opaque, and a flocculent precipitate separates out on letting it stand for several hours in the cold. The precipitate from 25 c.c. peptone-plasma, when washed and dried, amounted to 0·26 grm.[1] It has the properties above enumerated—i.e. it contains a quantity of lecithin which may be extracted with alcohol, as well as an iron compound which is not affected by hydrochloric acid and alcohol, and so forth. The experiment gave the same result when repeated with other portions of peptone-plasma.

[1] I may remark that peptone-plasma contains about 2 per cent. of fibrinogen (dry weight).

3. A fibrinogen is precipitated from peptone-plasma by cooling, and dissolves slowly in 0·2 per cent. HCl. A considerable precipitate arises after pepsin has been added and the mixture has stood for several hours in the incubator.

4. All the fibrinogen is precipitated from plasma by dilute sulphuric acid, and separated by means of the centrifuge. The precipitate is dissolved in an excess of acid, treated with pepsin, and allowed to digest, when again we get a precipitate.

5. All the fibrinogen is precipitated from peptone-plasma by means of sodium chloride. As I have elsewhere shown,[1] this body undergoes some change in this process, so that it will not now dissolve in dilute acids or in a normal saline solution. In this case it gives a perfectly clear fluid on digestion.

6. Coagulation was brought about in 300 c.c. of peptone-plasma by passing a current of CO_2 through it. The fibrin, when pressed out and washed, forms a perfectly clear solution on being subjected to artificial digestion.

7. Normal dog's serum and serum from peptone-plasma, after complete coagulation has taken place, yield no or scarcely any precipitate when artifically digested. The faint cloudiness that may arise is due to the fact that dog's serum usually contains traces of a fibrinogen.

8. Some dissolved tissue-fibrinogen (from thymus) is added to very strongly peptonised plasma, and coagulation thereby set up. As soon as this is evident, the mixture is stirred with a fine glass rod for the purpose of drawing out the fibrin in threads. These threads, collected and digested, give rise to a perfectly clear solution. A further addition of a small quantity of tissue-fibrinogen produces a fresh crop of coagulum, which may be repeated even a third time. This observation shows that very complex processes are at work, and the following results tend to the same opinion.

As the second coagulation cannot be adequately explained as being due to the further addition of a very small amount of tissue-fibrinogen, it is apparent that coagulable matters must

[1] 'Uebersicht einer Theorie,' &c., Ludwig's *Festschrift*, 1887. [Collected Papers, p. 186.]

have remained over from the first coagulation. This is, in fact, the case; for the serum of the first clotting yields a precipitate on the addition of dilute mineral acids, and gives rise to an insoluble residue when digested. Fibrinogen is therefore still present.

It may be proved that this fibrinogen still present cannot be referred to the original tissue-fibrinogen, for although a precipitate is formed by acetic acid in the fluid obtained after the first clot has separated, yet the precipitate is again dissolved in an excess of this acid; whereas, if tissue-fibrinogen be precipitated from its solution with acetic acid, the precipitate does not disappear on adding an excess. Moreover, the serum of the first coagulum, when added to fresh peptone-plasma, does not produce clotting, which would infallibly take place were tissue-fibrinogen present.

This residue of fibrinogen is, however, not identical with the fibrinogen of the original peptone-plasma, for the serum from the first clot, upon which the tissue-fibrinogen proved so effective, cannot be coagulated by dilution and treatment with carbonic acid, in spite of the presence of fibrinogen, whereas fresh and strongly peptonised plasma invariably clots by either means, and more rapidly if ferment be present.

If the same experiment be made with a weakly peptonised plasma, nearly the whole fibrinogen of the plasma passes into the coagulum on the first addition of tissue-fibrinogen. Mere traces remain, which I have designated as 'serum-fibrinogen.'

By strong peptonisation, therefore, the fibrinogens of the plasma may be rendered highly resistant, so that coagulation only occurs by degrees when tissue-fibrinogen is added. The portions of fibrinogen which do not coagulate the first time must, however, always, as observation shows, undergo alterations. It is, indeed, rendered obvious from the simple fact that they are included in the coagulum on further addition of tissue-fibrinogen.

II. *The Significance of Precipitation for the Process of Coagulation*

The fibrinogens behave very differently towards the various substances which bring about coagulation. The least altered fibrinogens of the blood-plasma are not affected by fibrin ferment, but, as I shall proceed to show, require a second fibrinogen for clotting to take place. Other solutions of fibrinogen, such as the B-fibrinogen, which has been changed by a single precipitation, clot not only with other fibrinogens, but also with ferment. Many hydrocele fluids act in the same way. There are also transudations which coagulate readily with ferment, but not at all or very slightly with fibrinogen. It appears that the fibrinogen of the plasma is altered during its passage through the walls of the blood-vessels, or as a result of standing after it has left the vessels.

It may here be remarked that serum is not a suitable reagent to be used as a comparative test for different solutions of fibrinogen. Ordinary dog's serum contains two constituents which are capable of initiating coagulation in solutions of fibrinogen. One is fibrin ferment, which, according to Hammarsten, is not a proteid, and which produces clotting very abundantly in dilute magnesium sulphate plasma. It is quite certain that it is destroyed on heating, and that it does not itself enter into the formation of the clot.

The other constituent is serum-fibrinogen, a body which, like all fibrinogens, is thrown down from its solutions by acids.[1] Like these, also, it consists of proteid and lecithin, and its action is quite different from that of fibrin ferment. It causes peptone-plasma to coagulate, which ferment cannot do. In this coagulation it disappears from the plasma, and the amount of fibrin formed is equivalent to the quantity of serum-fibrinogen added. Unlike fibrin ferment, it is without influence on magnesium sulphate plasma. Serum-fibrinogen is only present in dog's

---

[1] 'Note on a New Constituent of Blood Serum,' *Proc. Roy. Soc.* March 31, 1887. A similar substance may be prepared from ordinary fibrin. [Collected Papers, p. 169.]

serum in very small amounts. If it be precipitated with acid, and collected from large quantities of serum, and then dissolved and injected into the circulation of a rabbit, the blood drawn off from a vein after the injection will remain fluid for several hours; but if, on the other hand, the original dog's serum, which contains mere traces of serum-fibrinogen and a large proportion of ferment, be injected, no visible effect is produced either on the animal or on the blood. I am, therefore, disposed to consider the so-called ferment intoxication as a fibrinogen action.

When coagulation occurs in solutions of fibrinogen as the result of the interaction of two fibrinogens, the precipitation of one of the two bodies appears to be a necessary preliminary condition. Peptone-plasma contains two fibrinogens which I have termed ' A- ' and ' B-fibrinogen,' and which readily act upon each other to form fibrin. But they have little influence on each other until the A-fibrinogen is precipitated from the solution by some means. If peptone-plasma be kept at 37° C., and carbonic acid passed through it, no clotting occurs ; while this takes place if the plasma be kept at the temperature of the room, and still more rapidly if the plasma has been previously cooled, so as to start the precipitation of the A-fibrinogen.

The effect of a single precipitation is strikingly shown by the following experiment :—The fibrinogen is separated from a portion of peptone-plasma by a strong solution of common salt; the precipitate is filtered off and dissolved in dilute salt solution, and then added to another portion of the same plasma. Coagulation rapidly sets in. As peptone-plasma does not contain any ferment, the occurrence cannot be regarded as the result of a fermentative action on the precipitated and redissolved fibrinogen. But if it be remembered that, as I have shown in detail in the ' Uebersicht,' the fibrinogens of the plasma are altered by precipitation, it will be understood how it is that the fibrinogen, after it has been once thrown down, behaves like a foreign substance on admixture with the original plasma, and acts exactly as if tissue-fibrinogen had been added.

The nature of the change which the fibrinogens undergo

through precipitation has not so far been ascertained. But the facts that all fibrinogens are rich in lecithin, and that the latter plays an important part in coagulation, would tend to show that the amount of lecithin in the substance is altered or that a molecular rearrangement has taken place. The following experiment seems to show that this view is correct. In preparing fibrinogen from dog's plasma it frequently happens that the first precipitate is so much changed that it resembles fibrin. In this state it can only be redissolved in dilute salt solutions with difficulty, and scarcely becomes cloudy when artificially digested; just as in fibrin digestion there is no appearance of the abundant precipitation of any body resembling nuclein. If it be taken into consideration that the phosphorus contained in the residue which is left undissolved from the digestion of the fibrinogens is almost, if not entirely, to be referred to the lecithin, it follows that the mutual relations of proteid and lecithin in the fibrinogen must have undergone such considerable alterations in consequence of the precipitation that it is no longer possible to split up the tough nuclein-like product. The lecithin or the constituents containing lecithin are probably thereby enabled to act more readily upon neighbouring fibrinogens. Similar theories with regard to the influence of fibrin ferment on the altered fibrinogens will readily occur in this connection.

III. *The Action of Tissue-fibrinogen on Circulating Blood*

If solutions of tissue-fibrinogen be injected into the circulation of a living dog intravascular clotting takes place;[1] but, curiously enough, it only occurs in definite parts of the vascular system. If the experiment be made upon a fasting animal, or upon one that has been fed on lean meat only, and if the solution be allowed to flow from the jugular vein into the right cavity of the heart, thrombosis is usually brought about only within the region of the portal vein, and is the more widespread the greater the amount of tissue-fibrinogen injected. But it is

[1] *Proc. Roy. Soc.* 1886. ' On Hæmorrhagic Infarct of the Liver,' *Lancet*, Nov. 8, 1887, and *Brit. Med. Journ.* Nov. 8, 1887. [Collected Papers, pp. 135, 346.]

difficult to produce coagulation in other parts of the vascular system, however large a quantity be employed.

On the other hand, if well-fed animals in full digestion be used for the experiment, clotting will occur in the right side of the heart and in the pulmonary artery; and rapid injection may cause such complete thrombosis of the right side of the heart as to interrupt the circulation and to cause instant death before the solution can reach the portal vein at all.

The fact that in fasting animals the tissue-fibrinogen passes through the right side of the heart, the pulmonary circulation, the left side of the heart, and finally through the intestine, without causing any disturbance, and that coagulation does not begin until the portal vein is reached, shows undoubtedly that the blood must acquire a certain character in the intestine which it again loses in its passage through the liver. The following observation will, I think, indicate the nature of this alteration.

If blood be taken from the animal shortly after the injection of tissue-fibrinogen, it will be found to coagulate excessively slowly, and clotting may be delayed for hours or for days, according to the amount of tissue-fibrinogen injected. The plasma nevertheless contains a considerable quantity of fibrinogen, which can be made to coagulate by the addition of lecithin or tissue-fibrinogen; on the other hand, it is proof against the methods which initiate clotting in cooled plasma or ordinary peptone-plasma—such as filtration through a clay cell, dilution, treatment with CO_2. In fact, it behaves exactly like a plasma from which a substance important in the formation of fibrin—A fibrinogen—has been removed, as I have described in my 'Uebersicht.' It thus appears that the introduction of tissue-fibrinogen into the circulation has caused the disappearance of this body from the blood; and, having regard to its great coagulability, it is probable that it unites with the tissue-fibrinogen to form the thrombi in the portal system. It would also follow that the blood in the portal vein is richer in A-fibrinogen than that in other vessels.

We may adduce other observations in support of this

conclusion.   In a previous paper[1] I mentioned the fact that the amount of A-fibrinogen in the blood is increased by a diet of fat meat, and I have already mentioned in this paper that intravascular clotting also takes place in the right side of the heart, and in the pulmonary artery of animals which have been fed on fat.   When the blood has been flooded with A-fibrinogen by a copious diet of fat, this substance appears to be no longer confined to the portal vein, either because it has passed through the liver or because it has flowed through the thoracic duct. Thrombosis of the portal vein, after the injection of large quantities of tissue-fibrinogen, causes such an immediate and considerable fall in blood-pressure that it is difficult to obtain any blood from the carotid.   In spite of this the animals usually recover.   Some of the clots disappear, but others lead to pathological changes in the liver, such as hæmorrhagic infarcts, followed by fatty degeneration and formation of connective tissue.

I hope shortly to give a full account of these changes.

The action of tissue-fibrinogen on the circulation of the dog is as invariable as that of peptone.   In thirty experiments I obtained thrombi in the portal vein without a failure.   There is also a certain analogy between the action of peptone and that of tissue-fibrinogen.   Both inhibit coagulation, both act quantitatively, and chiefly on the portal system.   Peptone paralyses the intestinal vessels, tissue-fibrinogen coagulates the blood in them.

[1] Cf. Pawlow's interesting observations on the changes which the blood undergoes in the lungs, *Du Bois' Archiv*, 1887.

*SECOND SECTION OF REPORT TO THE SCIENTIFIC
COMMITTEE OF THE GROCERS' COMPANY*[1]

Note by Editors

[THE order of subjects treated in this second Section of Report is as follows :—

I. The significance of proteid modifications.

II. The separation of (*i.e.* possibility of separating) different proteids from proteid solutions.

III. Chemical identity and individuality as applied to proteids.

IV. The properties of A-fibrinogen, and complex composition of fibrinogens.

V. The properties of stroma and tissue-fibrinogens.

VI. The properties of plasma-fibrinogen.

VII. The relation between lecithin and proteid.

VIII. On the processes termed 'coagulation,' with especial regard to the *rôle* played by lecithin.

IX. Quantitative relationships in fibrinogen interactions.]

I. *The Significance of Proteid Modifications*

The fibrinogens form an important class of proteid substances very closely allied, as I have pointed out, to the casein of milk. On the other hand, they have many characters which have led to the belief that they belong to the group of proteids known as

---

[1] Dated April, 1889. [Dr. Wooldridge intended that the Report of his researches to the Scientific Committee of the Grocers' Company should comprise five sections. Of these, only the first was printed. He had however drawn up a rough draft of the second section, which is reproduced here, but had not completed it at the time of his death. For the order of sequence of the various subsections in this paper, the editors are therefore responsible.—ED.]

'globulins.' I think it unfortunate that they should be classed as globulins. We look upon the globulins as a special modification of proteid—*i.e.* a particular form which is insoluble in water, soluble, however, in neutral salt solutions. The whole modern classification of proteids, which is obviously based on a number of utterly unproven assumptions, is to my mind a little dubious; and it is not clear to me that the *so-called* varieties of proteid mean what they seem to. It is essential that I should briefly allude to the view I take of the subject, as otherwise my position with regard to coagulation cannot be understood.

There have been obtained crystallised proteids.[1] These proteids have been obtained so free from ash that the solution is of itself an antiseptic solution (Drechsel)—*i.e.* the mere solution exposed to air will not serve as a soil for the growth of bacilli. Even large quantities give but the merest trace of phosphorus reaction when incinerated. Phosphorus in traces is always found, and probably belongs to the salts which so constantly accompany proteid matter. Whether these salts are in genuine chemical union or not with the proteid seems at present undecided. These proteid crystals (*Paranuss*) are probably the purest form of proteid ever obtained; they are 'globulins,' they are not soluble in water.

Now it is very much open to question whether we have ever obtained any other proteid in an approximately so pure a state as these crystals. In any classification of proteids, then, in which the separation of proteids into groups is made to depend upon conditions of solubility, it is obviously very easy to make serious errors. It is customary, for instance, to speak of 'native albumens,' a very important character of which is that they are soluble in water. And in speaking of *native albumens* and *globulins* we imply that they are both *proteid modifications*. But we do not at all know on what the solubility in water of the 'native albumen' depends. It has never been obtained in a condition of purity approximately like that of the crystals, and,

[1] Schmiedeberg, *Zeitschr. f. phys. Chem.* 1877, and E. Drechsel, *Journ. f. prakt. Chem.* N.F. vol. xix. p. 331.

for aught we know to the contrary, its solubility in water may depend on an admixture. The 'globulin' may be the simplest expression of proteid. Now we are all agreed in separating the casein of milk sharply from native albumens, and this is certainly not a mere proteid 'modification.' It is, by general consent, a highly complex phosphorus-containing substance. Yet casein has just as much ground to be considered soluble in water as native albumens have. We can dilute neutral milk to any extent with water ; the casein is not precipitated. To assure one that in this case something associated with the casein is the cause of its apparent solubility in water, and to make no such assumption in the case of serum albumen or egg albumen, is clearly quite illogical. Equally illogical is the supposition that, because a precipitate obtained from milk by an acid requires alkali for its solution, therefore the original casein also required that alkali. It involves the assumption that the precipitation is an indifferent process. Now it can be shown in the most striking manner that precipitation of proteids, even by very weak acids, is anything but an indifferent process.

To return to the fibrinogens : as they occur in nature .they are *soluble in water* just as the casein of milk is soluble in water. The fibrinogens always occur with *admixtures*, and these admixtures may be the cause of their solubility in water. It is doubtful whether any proteids form a true solution. Like casein, the fibrinogen 'solutions' are more obviously 'Aufquellungen' than other proteids. Thus milk will filter through paper but not through a clay cell. The fat particles and the fine pores together stop its progress. This is also the case in a marked manner with regard to the fibrinogen of the blood. As I have long pointed out, no trace of fibrinogen will pass through a clay cell if the plasma be fatty.

II. *The Separation of different Proteids from Proteid Solutions*

It is essential that we should consider this question in its general aspects, as it has a most important bearing on the chemical side of coagulation. It will be convenient to consider the behaviour of plasma, as it illustrates the points at issue very clearly.

It is easy, by treating plasma in different ways, to make it apparently quite evident that it consists of a number of different varieties of proteid.

We may first consider the question of heat-coagulation as an indication of the existence of a special proteid. In no case can this method of identification be used with more apparently conclusive results than in the case of the plasma. Certain forms of plasma give a precipitate when heated to 55°–56°. From certain forms of plasma a proteid can be precipitated by other means, and its solutions in certain strength of salt also coagulate at 55°–56°. After the plasma has been heated to 55°–56°, and the precipitate produced, it will not yield fibrin. On the other hand, the proteid, which, isolated from the plasma, coagulates at 55°, will yield fibrin. This appears very conclusive evidence that the plasma contains a special proteid, that we can identify this proteid, as it occurs in plasma, by means of its heat-coagulation temperature, and that this proteid is the body from which fibrin is produced.

But we are met at once by a great difficulty. As I have pointed out, all forms of plasma do not coagulate at 56°—*e.g.* the 5 per cent. NaCl plasma of the dog. Nevertheless it can be made to yield fibrin and to yield a proteid solution coagulating at 55°–56°.

The easiest way out of this difficulty is to make an assumption ; to suppose that there is something or other which interferes with the coagulation of the particular proteid at 55°–56°. It is easy to make such an assumption plausible, but very difficult to show that it actually exists in the case before us. (As a matter of fact, in the particular case before us it can be shown to be a quite wrong assumption.) And once we have to begin to make assumptions of this sort we are practically admitting that the method of identifying proteids by their heat-coagulation temperature is an extremely unreliable procedure.

Again, the arguments based on experiments with the salt precipitate from salt-plasma present equal difficulties.

For many years it has been known that various forms of salt-plasma yield, on saturating with neutral salt, a proteid precipitate. Denis regarded this precipitate as one body, *plasmine*, which he thought split up into ' fibrine concrète ' and ' fibrine dissoute ' on coagulating. Alexander Schmidt, on indirect grounds, thought it consisted of two bodies which united to form fibrin. Hammarsten, again, supports Denis.

But it is clear, from the past history of the question, that with the data made use of by these investigators the question cannot be decided. It must always remain a matter of opinion, each observer supporting his own opinion by observations which, whilst making it plausible, do not prove it.

The coagulable matter of the plasma.—For many years it has been known that it is possible to isolate from plasma a proteid which is called ' fibrinogen.' This proteid is capable of forming fibrin ; as regards fibrinous coagulation, its solutions resemble in some respects plasmata. Whenever coagulation has occurred in plasma this proteid has disappeared to a greater or less extent ; and when this proteid has been removed from plasma it is no longer capable of fibrinous coagulation.

Hence it is customary, in dealing with the chemical aspects of the coagulation question, to assume that the substance of plasma chiefly or wholly concerned in coagulation is fibrinogen ; and doubtless there is some truth in this assumption.

By particular means we can separate from plasma a substance which is more immediately concerned in fibrin formation than any other constituent of the plasma. But we make a most unwarrantable assumption if we suppose that, when coagulation occurs in *plasma*, this substance being present, *the chemical changes are limited to this substance.*

I illustrate this by an example. The fibrinogen obtained from blood-plasma interacts with a solution of tissue-fibrinogen, an obvious result being produced—coagulation.

Again, tissue-fibrinogen interacts with plasma, an obvious result being produced—coagulation.

But we cannot assume that in the two cases the chemical process has been identical, since we have positive evidence that when the tissue-fibrinogen acts on plasma or blood there are profound changes produced in the plasma and the blood quite outside the immediate sphere of coagulation.[1]

If we isolate from plasma a fibrinogen, and study its behaviour when coagulation is artificially produced, and apply the results so obtained to the phenomena of normal coagulation, our conclusion may be correct or it may be quite misleading, because we have to make a series of assumptions as follows :—

1. The process of separation alters neither the substance precipitated nor the plasma left.

2. The substance isolated treated in a certain way (a) undergoes certain changes.

3. Therefore, when this substance is present in a highly complex fluid, and is treated by method (a), the chemical changes are limited to the fibrinogen, do not affect the other constituents of the plasma, and are of the same character as those which occur in the isolated solution.

Now clearly these assumptions are anything but satisfactorily proven, and hence the coagulation question, considered as a chemical problem, presents such insuperable difficulties. To ignore these difficulties is to retard true progress.

III. *Chemical Identity and Individuality as applied to Proteids*

Let us suppose we have a fluid containing proteid in solution. By a certain procedure we separate a part of the proteid. To what extent are we entitled to speak of this proteid as a substance or chemical individual ? I will illustrate the difficulties of this question in its bearing on coagulation. I have already alluded to them in speaking of the separation of different proteids from plasma.

In dealing with the substances called ' tissue-fibrinogens ' I obtain, by means of decided acidification, a proteid precipitate, and I speak of this substance as if it were a chemical individual. Now it would be very easy to manufacture out of this proteid, or out of the solution from which it has been obtained, a great

[1] *Vide* ' Auto-infection in Cardiac Disease,' *Proc. Roy. Soc.* 1888, p. 309. [Collected Papers, p. 350.]

number of varieties of proteid, if heat-coagulation could be relied on as an indication of different forms of proteid.

According as the solution is very fresh or a little older, according to the concentration of the solution, according to the presence or absence of traces of salts, its behaviour towards heat may be made to vary enormously. It may sometimes be boiled and no coagulum separate at all, or a certain amount of coagulum may be formed whilst the bulk remains in solution, or the bulk may coagulate and traces or nothing of the proteid remain in solution. And by heating to temperature lower than boiling a still greater variation may be obtained.

So, too, in dealing with the precipitate produced by acid, it can easily be apparently separated into different substances, especially if it have stood under water for some time. Thus, on treating it with 5·0 per cent. solution of sodium chloride, a part of it goes into apparent solution, another part is dissolved, if at all, with great difficulty. This might be taken as an indication that there were two substances in the precipitate. But, inasmuch as the fresher the precipitate is the more easily it goes into solution, and also the greater the amount of salt solution used the more goes into apparent solution, the meaning of such a separation becomes dubious.

In my first work on the blood I have described a method of preparing the stromata of the red corpuscles. The method depends on acidification. At that time (1880) I was unacquainted with the fibrinogens or their properties; I was also ignorant of the relation of the blood-corpuscle stromata to coagulation. I at that time described the stromata as consisting of paraglobulin and a substance more or less like ' nucleoalbumin.' But, as I pointed out in that paper, there were great difficulties in assuming that these bodies existed side by side. One had to make suppositions. Since my observations on tissue-fibrinogen and the relation of the stromata to coagulation, I speak of the stromata as chiefly consisting of fibrinogen, and I regard my previous separations as artificial.[1]

---

[1] *Vide* ' Relation of Red Corpuscles to Coagulation'; also ' Blood-plasma as Protoplasma.' [Collected Papers, pp. 167, 172.]

Now I must point out the reason I have for adopting this position, and for supposing that both tissue-fibrinogen and stromata are chemical individuals. I think the basis of all our views on the chemistry of coagulation is formed by my observations on the body I have termed ' A-fibrinogen.'

This substance is precipitated from peptone-plasma by simple cooling. With moderate precautions in dealing with peptonisation its properties are remarkably constant, and I cannot avoid the belief that its peculiar morphological appearance is an indication of crystallisation, because it so closely resembles the early stages of crystallisation in the crystalline proteids.

Now I am strictly of the opinion that in dealing with complex proteid bodies crystallisation is an absolute desideratum. Where would our chemical views of hæmoglobin be if we had not got it crystallised ? We cannot possibly decide whether we are dealing with an admixture or with one substance until crystallisation is achieved. There will be endless views, all more or less plausible, all doubtful.

To maul a proteid mass about until it becomes indifferent to further mauling, and then to make an elementary analysis and suppose that thereby any step has been taken towards elucidating the chemical processes of the organism, is the most insane idea that ever occupied the mind of man. The results obtained are absolutely grotesque, and they are excessively injurious, for they throw over the whole subject an air of exactitude, whereas in reality they entirely obscure it. As an illustration I may cite Hammarsten's analysis.

This observer is of opinion that fibrinogen will pass into the coagulated modification under two conditions—the action of the ferment and the action of heat (temperature of 55°). Under both of these conditions Hammarsten finds that a part of the proteid may remain in solution.

For the use of the action of the ferment, Hammarsten would fain prove that there is a dissociation of the fibrinogen into two bodies—fibrin and a globulin ; and to strengthen this view he has given elementary analyses of the fibrinogen, the fibrin, and the globulin. There is a marked difference between the

globulin and the fibrinogen and fibrin. But it is obvious, in
the face of Hammarsten's own descriptions, that these elementary
analyses are entirely useless. Hammarsten says that, however
attractive this hypothesis may be as an explanation of the action
of the ferment, the analogy of casein being strongly in his mind,
there exist the greatest difficulties in the way of its acceptation,
because actually the ratio of fibrin to the dissolved proteid is
subject to the greatest variation. Now if we consider the
'elementary analyses' under the light of this admission, we see
that a proteid, A, splits into two others, B and C, which have
different compositions. A may split into $\frac{2}{3}$ B and $\frac{1}{3}$ C, or $\frac{1}{3}$ B
and $\frac{2}{3}$ C, &c. Under these circumstances what possible light
can elementary analysis throw on the question, since it is certain
either that the analyses give no information at all, or that A or
B or C are bodies of very varying composition?

IV. *The Properties of A-fibrinogen, and complex Compositions
of Fibrinogens*

A distinguished and clear-headed physiological chemist,
who looks on the subject from the chemical aspect, says 'we do
not know what constituents of the plasma have taken part in
coagulation.'[1] When he wrote that it was perfectly true.

The discovery of A-fibrinogen, its behaviour towards coagulation, and its properties, fundamentally alter our views of the
chemistry of coagulation. The following considerations will,
I think, justify this statement.

1. The fact that mere slight cooling suffices to precipitate it.

The danger we have to guard against in this kind of investigation is to avoid unwarrantable assumptions, to avoid artificial
results. If we apply some potent reagents to plasma, such as
an acid, or saturate it with a salt, we are exposing ourselves to
great fallacies. We do not know what these reagents do to the
plasma or to the precipitate, and we get into endless difficulties
with the precipitate. Is it one, two, three different substances?
Tot homines, tot sententiæ.

Now I make the assumption that slight cooling as a method

[1] Bunge, *Physiological Chemistry*, p. 241.

of separating a substance is the least likely to produce profound disassociations in the plasma, and is the least likely to profoundly alter the substance obtained. I shall quote the grounds I have for this later; *a priori* it seems not wholly unwarrantable.

2. Its morphological characters.

As I have frequently pointed out, it separates in regular disc-like granules. To study and appreciate these granules properly, slight and very slow cooling is necessary. It must be remembered, of course, that we are examining a substance which, for a certain time, is easily re-soluble in the fluid in which it is examined as soon as the temperature goes up again, and the first result of warming is to make the individual granules swell up, fuse together, and form large spheres, which swell and swell until they become lost in the fluid, when re-solution has taken place.

Method of examination of plasma.—The quickest way to procure A-fibrinogen is to cool the plasma by placing the vessel containing it in ice. In a few minutes a precipitate is produced, which, if examined immediately, is seen to consist of extremely small regular granules; if examined later, it is found to consist of discs and spheres of varying size, shape, and form—in fact, it is in the stage of re-solution.

To fully appreciate the close similarity of this precipitation to an incomplete crystallisation the following precautions must be observed: The plasma must be freed as quickly as possible from all form-elements by repeated centrifugalising; the plasma is then put in *flat* (*vide* p. 288) *dishes in moderate quantity*; the dishes are covered, and it is allowed to cool very slowly. A convenient way is to leave it simply on the laboratory bench. The laboratory becomes slowly cooler overnight, and the precipitate must be examined before the room gets appreciably warmer. The plasma is then full of regular, pale round discs of great uniformity of size and appearance, generally about half the size of a red corpuscle. If there have been any rewarming, a few larger or compound discs or globules may be seen. Rapid cooling then produces very minute regular granules,

T

slow cooling large uniform regular granules. If, then, we consider these facts, and also remember that in the crystallised proteid the first step towards crystallisation is the appearance of regular rounded granules, it seems to me not an unwarrantable conclusion that this precipitate represents a chemical individual. For I cannot conceive what this curious form of precipitation can mean if it does not mean an imperfect crystallisation.

In discussing this subject it must be remembered that A-fibrinogen separates in a fluid which is by no means an indifferent one; that is to say, as soon as it is separated it becomes slightly altered and tends to interact with the fluid plasma. This is especially the case with slightly peptonised plasma, in which the precipitate of A-fibrinogen rapidly becomes converted into fibrin. Favouring conditions will be pointed out later. To obtain, therefore, uniform results the plasma must be moderately strongly peptonised. (*See* later.)

(a) *The alteration of A-fibrinogen on precipitation.*—For some time after its precipitation by cooling, A-fibrinogen retains its power of redissolving in the plasma on warming to rather above its original temperature. But after a time it loses this power; that is to say, it undergoes an 'alteration' consequent on its precipitation. In the case of A-fibrinogen this alteration is a most interesting but a complex process. I mean, it is difficult to know how much is due to intramolecular changes of A-fibrinogen itself, and how much to interaction between it and the surrounding fluid. From the experiments I described in Section I.[1] (p. 244) it would seem highly probable that it undergoes an alteration directly it separates out.

The re-solution of A-fibrinogen may be effected not only by warming the plasma, but by the addition of sodium chloride solutions (to make whole 10 per cent.), or by adding weak solutions of alkali—caustic or carbonate. The effect of the 'alteration' is here also very apparent. After prolonged separation the neutral salt causes it to swell and imperfectly dissolve, or

---

[1] [By Section I. is invariably meant the First Report to the Scientific Committee of the Grocers' Company, p. 201.]

merely to swell. Dilute alkalies have the same action, though the change must be more pronounced.

A-fibrinogen in its native conditions is soluble in water. If peptone-plasma be diluted ten times and a current of CO_2 passed through it, a precipitate is produced which has none of the properties of fibrinogen and, so far as can be ascertained, no influence whatever on coagulation; this is paraglobulin. After it has been removed the plasma can be diluted to a very large extent—one hundredfold—without the occurrence of any further precipitate. A very slow coagulation may occur, but no precipitate whatever within a moderate time—hours.

(b) *Action of acids on A-fibrinogen.*—A-fibrinogen, when collected by centrifugalising from moderate peptone-plasma, behaves thus with regard to acids :—With dilute HCl (0·2 per cent.) it dissolves; although there is nearly always an almost imperceptible residue. It does the same with dilute sulphuric acid, but not quite so readily. With acetic acid it becomes opaque ; with strong solutions (5 per cent. and more) some of it may dissolve.

A-fibrinogen may be precipitated from the plasma—

1. By dilute acids.

2. By saturation with neutral salts. (*Vide* Plasma-fibrinogen (p. 281).)

(c) *Behaviour of A-fibrinogen towards artificial digestion.*— If to the HCl solution of A-fibrinogen pepsin be added, and digestion allowed to go on, a flocculent precipitate makes its appearance. The precipitate is soluble in alkalies, gives modified proteid reactions, and is extremely rich in phosphorus. The phosphorus is in the form of lecithin.[1]

A-fibrinogen itself always contains ash ; I have not investigated it more exactly. It also always contains iron, and the iron is in the digestion precipitate.

The following properties of A-fibrinogen have been enumerated :—

1. Separation in definite rounded granules.

2. Alteration on separation.

---

[1] For particulars with regard to lecithin, &c., *see* pp. 279, 282, 284, 290.

3. Solubility in water.
4. Precipitation by acids.
5. Precipitation by neutral salts.
6. Precipitation by digestion.
7. Chemically A-fibrinogen consists of proteids and lecithin, ash (unknown composition), traces of iron. There are perhaps other bodies—*e.g.* cholesterin. I have only been able to recognise two substances distinctly—proteid and lecithin.

To these must be added two other properties, which I discuss later.

8. It enters materially into the formation of fibrin.
9. It has the power of inducing the appearance of fibrin.

In all essential points the tissue-fibrinogens, the lymph-fibrinogen, the stromata of the red corpuscles agree with A-fibrinogen, and for this reason I am strongly of opinion that we must regard all these bodies as complex chemical individuals.

As I have pointed out, the phenomenon on which I think more weight is to be laid than any other in regarding A-fibrinogen as a chemical individual is its pseudo-crystallisation. As I have explained, special conditions are necessary to obtain what may be called a good and typical coagulation. A-fibrinogen may be precipitated by other means than cooling, and may occur, when so precipitated, in rounded granules, but they are not so characteristic nor so large. In dealing with tissue-fibrinogen and lymph-fibrinogen I have not been able to get them in distinct large granules. From the regularity of the granules obtained under special conditions, I think it a most promising substance to obtain in a crystalline form, and I have carried out much research in this direction. But at present I am obliged to adopt this position :—

A-fibrinogen there is good reason to regard as a chemical individual. Tissue-fibrinogen resembles in all points A-fibrinogen, except that we cannot get it in definite pseudo-crystals; therefore I conclude that tissue-fibrinogen is a complex chemical individual. It would be easy to artificially split it up into great varieties of proteids; but for the reasons I have adduced, and those I shall adduce, I think my view is a correct one.

I am sure that the one thing which can put our knowledge of coagulation on any exact chemical basis must be crystallisation. It becomes more and more clear that all the work which is done on complex proteid substances, unless it be done in the direction of crystallisation, is provisional, is dubious, and does not come into the domain of chemistry (as the chemists know it). As I have already hinted, the classification of proteids into 'modifications' rests on dubious grounds. I shall afterwards consider this.

V. *The Properties of Stroma- and Tissue-Fibrinogens*

The stromata of the red corpuscles.—As I have already mentioned, the stromata of the red corpuscles act in every way as fibrinogens, their action on intravascular and extravascular blood being quite similar to that of tissue-fibrinogens. The conditions of their precipitation (acid) are quite similar to those of the tissue-fibrinogens, and their general chemical behaviour is similar. It is, as in the case of the fibrinogens, quite easy to make apparent separations into different substances, but, from reasons stated above, it is highly probable that these separations are artificial, and I think this is especially the case in dealing with the red corpuscle stromata. I am much inclined to think that the stromata are a discoidal precipitation of fibrinogen, and not, as is generally thought, 'a protoplasmic residue' of form-elements.

The importance of the close similarity between A-fibrinogen and the stromata and the possible conclusions which may be drawn therefrom, I have already discussed in my Arris and Gale lectures, and the fact that the stromata are to be regarded in the main as fibrinogen I have on several occasions pointed out.[1]

These views are, of course, quite at variance with those

---

[1] *Vide* 'Relation of Red Corpuscles to Coagulation,' 1886. 'Beiträge zur Frage der Gerinnung,' *Du Bois' Archiv*, 1888. 'Blood-plasma as Protoplasma,' 1886. [Collected Papers, pp. 167, 253, 172.]

expressed in my first paper in ' Du Bois.' [1] But at that time
the relationship of the red corpuscles to coagulation and the
nature of fibrinogen were quite unknown, and I had to adapt
my observations to the existing knowledge. In dealing with
the question of Anæmia and Blood-formation, I shall have to
return to this question.

Lymph- and thymus-fibrinogens.—Though not precisely identical, the simple ½ per cent. NaCl fluid of lymphatic or of
the thymus glands, and the watery extract of the same glands,
contain essentially similar fibrinogens as regards coagulation
and main chemical details. The water extraction undoubtedly
exercises an influence, since it first causes a very transient
semi-precipitation or pseudo-coagulation in the slimy mass of
cells and fluid. This rapidly disappears, and the watery extract
contains, after a few hours, a very large quantity of tissue-
fibrinogen. This is largely derived, if the extraction be quickly
carried out, from the intermediate fluid. But the cells undoubtedly participate, especially if extraction be longer continued.

The watery extract can be enormously diluted without precipitation—*e.g.* 5 c.c. of a strong tissue-fibrinogen extract
diluted with 300 times its bulk of water gives not the faintest
precipitation, but immediately on adding acids a distinct precipitate is formed.

Precipitation.—The tissue-fibrinogen is precipitated by
rendering the fluid distinctly acid with very dilute mineral
acids or with organic acids ; mere faint acidification does not
precipitate. (This method of preparation excludes serum
globulin and paraglobulin. In its preparation not more acid
should be used than just suffices to precipitate the whole.)

After the removal of the tissue-fibrinogen the fluid contains
very small quantities of proteid matter, and it is totally without
influence on coagulation processes.

Tissue-fibrinogen can also be precipitated by saturation
with neutral salts. I am not conclusively sure that it can be
wholly precipitated by this means.

Alteration in solubility of precipitate.—Once precipitated

[1] [*Vide* Collected Papers, p. 69.]

the fibrinogen is 'altered.' Its 'solubility' in water is at once entirely lost or greatly diminished.

Freshly precipitated, it is more or less soluble in 10 per cent. salt solution. The longer it has stood the less soluble it is.

Freshly precipitated, it is very easily clearly soluble in dilute alkali, but after standing the solubility becomes impaired.

Freshly precipitated, it is soluble in excess of acid, HCl being the most effectual. The solution is never complete, and is mostly an 'Aufquellung.' It is very soon lost.

Great excess of acetic acid partially dissolves it if very fresh. So long as minute quantities of NaCl are present, solutions of tissue-fibrinogen are precipitated by acetic acid, and are not redissolved in very great excess.

Solutions of tissue-fibrinogen also 'alter' without precipitation; for every day the solution stands, the precipitate becomes much more insoluble in mineral acid, and entirely so in acetic acid (up to 20 per cent.). The solubility in alkali is retained for a longer time.

Behaviour on digestion.—If the swollen-up semi-solution in HCl (0·2-3 per cent. acid) be treated with pepsin, a precipitate is formed after a short time. The precipitate is soluble in alkalies, and gives proteid reactions. It contains as minimum 14 per cent. of its dry weight of lecithin. I am doubtful whether, after the complete removal of the lecithin, there is any phosphorus. Thus, in all its main characters, the precipitate of tissue-fibrinogen agrees with the A-fibrinogen.

The lecithin of tissue-fibrinogen.—To prepare lecithin from tissue-fibrinogen—*e.g.* for coagulation experiments—I adopt the following procedure. The washed precipitate is treated with strong alcohol, heated for a short time, then filtered, and the coagulum strongly pressed out. The first alcoholic extract I do not use, since it always contains proteid. The strongly pressed-out coagulum is then extracted with hot alcohol at 90°. To effect anything like complete removal many days' extraction is necessary. The alcoholic extract is evaporated on the water-bath. At first no attention need be paid to temperature; later, as the extract becomes concentrated, it should be moderately low. The extract is evaporated to complete dryness and again extracted with warm absolute alcohol in small quantity, the solution allowed to stand several hours in the cold, filtered, re-evaporated to complete dryness, exhausted with a small quantity of absolute ether, cooled, filtered, re-evaporated. This process may be repeated, for in my coagulation experiments I have carried out the alternate solutions in absolute alcohol and ether ten and twelve times. It is very wasteful and tedious.

The lecithin so obtained is a yellow waxy substance; it is quite insoluble in water or alkaline solutions, but it swells up and forms perfectly characteristic myelin drops. It is soluble with ease in alcohol, ether, petroleum, chloroform, benzol.

On incinerating imperfectly the ash is strongly acid; on incinerating with nitrate of potash and carbonate of soda the ash is found very rich in phosphoric acid.

When boiled with baryta water for two or three hours, the solution filtered off is found to behave as follows :—

It is treated with a stream of carbonic acid to precipitate the baryta; filtered, and evaporated. During these processes a further slight precipitate of barium carbonate may occur, which is removed by filtration. The filtrate then remains quite clear. This is evaporated to dryness, and extracted with absolute alcohol. The solution gives a precipitate with platinum chloride. This precipitate is easily soluble in water, and by slow evaporation separates out as large prismatic crystals of the double salt of neurin and platinum chloride. After the extraction with alcohol the residue is dissolved in water and filtered, forming a perfectly clear solution. This gives a doubtful reaction for phosphoric acid when tested directly with the molybdate test. On reevaporating the solution and incinerating with soda and saltpetre the residue is obviously organic, as it blackens and burns. The ash is very rich in phosphorus and contains barium (glycerine-phosphate of barium). The barium is estimated directly by precipitation with sulphuric acid.

I have not carried out the identification of the decomposition-products further. I am still engaged on them. But, if we consider the solubilities of the substance and its behaviour towards baryta, it is difficult to conceive of its being anything else but lecithin.

The original substance gives a flocculent bulky precipitate with $PtCl_4$. This is quite insoluble in water, readily soluble in chloroform, but with great difficulty in ether. It contains phosphorus, platinum, and chlorine.

Other constituents of tissue-fibrinogen. — In addition to lecithin and proteid, thymus-fibrinogen contains ash. It is easy to separate by weak HCl considerable quantities of magnesium and calcium phosphate. The ash of tissue-fibrinogen always contains iron. Whether the alcoholic extract of tissue-fibrinogen contains other substances than lecithin I am not quite sure.

Testis-fibrinogen.—The fibrinogen of testis agrees closely with that of thymus. (It has long been known that from the semen of rodents a fibrinogen could be prepared agreeing in most respects with the fibrinogen of Hammarsten, and, like this, yielding fibrin with ferment (Landwehr).)

The fibrinogen of testis is largely present in the intermediate fluid of the organ. In water its only apparent ' solubility ' is more marked than that of the thymus, so that mere filtration through paper, as soon as the pores get blocked with the *débris*, suffices to prevent any going through. An albuminous fluid passes through, giving no precipitate with acetic acid, and having not the slightest influence on the coagulation of plasma, intravascular or extravascular. The fibrinogen of testis is more easily dissolved in excess of acid, both acetic and hydrochloric, than are thymus- and lymph-fibrinogens.

It is exceedingly rich in lecithin, but I have not estimated the amount.

VI. *The Properties of Plasma-Fibrinogen*

The next form of fibrinogen I shall consider is *plasma-fibrinogen.* For purposes of general elucidation I have already considered A-fibrinogen. But it is not possible to obtain large quantities of this substance, since the actual weight of substance which separates from a considerable quantity of plasma is very small. To obtain chemical details I have had painfully to collect material, and store.

It is easier to get plasma-fibrinogen in larger quantity and in a pure state by other means.

As I have already pointed out, the NaCl precipitate from plasma can easily be regarded as two or one body. It is thought by one school of observers that it consists originally of paraglobulin and fibrinogen (Schmidt). Denis considered it one substance. Hammarsten considers it one substance which splits into two in the act of coagulation. Working as these observers have worked, the answer to the question would always remain a surmise.

I adopt the idea (for the present) that in dealing with plasma one can isolate three varieties of proteid : 1, fibrinogen ; 2, serum albumen ; 3, paraglobulin.

The method which I use of precipitating fibrinogen precludes a precipitation of either of the others, supposing them to be present as such. This method is acidification. I use dilute (four to six *pro mille*) sulphuric acid in preference ; and it is added very slowly, till the maximum of turbidity is obtained. No precipitate at all forms till there is decided acidity. Now dilute acids do not precipitate serum albumen, and paraglobulin is extremely easily dissolved in the slightest excess of mineral acid. Hence there is *a priori* reason for supposing that the acid precipitate is solely fibrinogen. But it is a very wasteful method. If sufficient acid be added, only a part of the fibrinogen is precipitated, if too much, it is easily redissolved ; and when dealing with the precipitate the sensitiveness to acid is still greater. The precipitation by acid is obviously aided by the presence of neutral salts, and hence I wash the precipitate with acidified ·5 per cent. NaCl, or very weak acid sodium sulphate. By means of the centrifuge the washing can be rapidly effected, but it is almost impossible to avoid a great loss in washing.

In the fibrinogen so isolated I found the following amount of lecithin : 4·1 grm. dry substance yielded 0·13 grm. lecithin. This determination gives a percentage of lecithin in plasma-fibrinogen of 3·2 per cent. It is certainly an under-estimate, because the substance still yielded traces of lecithin when I stopped extracting, and because it is impossible to avoid loss in the repeated solutions, filtrations, and re-evaporations which are quite unavoidable ; it is merely a minimal number.

I am still engaged on these quantitative investigations. But it is a serious undertaking to get much fibrinogen by the acid method. I use perfectly pure peptone-plasma which has been centrifugalised till it is as clear as water and perfectly free from the slightest trace of hæmoglobin.

The lecithin from fibrinogen prepared by the acid method presents all the characteristics of the lecithin prepared from tissue-fibrinogen. I have repeated with particular care the procedure with baryta water.

The substance was boiled with baryta for $1\frac{1}{2}$ hour, the filtrate treated with CO_2, and the baryta completely removed by filtration. The filtrate was evaporated to dryness ; one part was dissolved in water ; evaporated and incinerated it gave a very pronounced phosphorus reaction. The other part treated with alcohol gave a precipitate with platinum chloride. The insoluble residue from the baryta boiling was treated with dilute HCl. The previous heavy precipitate at once floated on the top. Shaken with ether, it at once went into solution (fatty acids). I have not been able to carry out further identification.

I have over and over again obtained lecithin from fibrinogen separated by the NaCl method. When I first found the lecithin action I spent at least two months in the endeavour to obtain fibrinogen free from lecithin. But it is quite impossible. More than three re-solutions and re-precipitations of fibrinogens from dog's blood I was never able to obtain. At each precipitation and re-solution a part became insoluble, until finally the precipitate would not dissolve at all any more, and the presence of lecithin in it could be demonstrated with perfect ease. The lecithin obtained from fibrinogen is most efficacious in producing coagulation.

The remaining properties of plasma-fibrinogen I will leave to the section termed 'coagulation processes.'

VII. *The Relation between Lecithin and Proteid*

Apart from the general reasons I have adduced above for regarding the fibrinogens as complex individuals, it cannot be decided whether the lecithin which occurs so abundantly in tissue-fibrinogen is chemically combined or not. Repeated solutions and re-precipitations do not in the least separate it. After the first or second precipitation all the organic matter is precipitated each time. The very great difficulty there is in separating the lecithin completely might be adduced as a reason for thinking it was chemically combined. But in reality we must look on the matter from a general standpoint. It has long been known that nearly every proteid precipitate from animal fluids is mixed with lecithin, and the relation of this lecithin to the proteid has been a matter of surmise. In some cases it has been supposed to be chemically united—*e.g.* the vitellin of egg; in others it has been supposed that the lecithin was mechanically carried down.

Further, in order to account for the clear *watery* solutions obviously rich in lecithin, it has been surmised that lecithin is dissolved by proteid, and it has also been supposed that the presence of lecithin alters the solubility of proteid. It has moreover been supposed that lecithin once in solution is easily carried down by proteid. Now all these surmises are totally lacking in any experimental basis whatever. If we prepare pure lecithin and try to dissolve it by means of albumen, we find our endeavour a conspicuous failure. I made a very large number of experiments to see whether I could get any evidence of albumen dissolving lecithin, but I could not ascertain that it had any power whatever in this direction.

Now it must be remembered that some of these solutions contain very large quantities of lecithin.[1] Thus, as I have stated, the digestion precipitate of tissue-fibrinogen contains over 12 per cent. The digestion precipitate of A-fibrinogen contains over 25 per cent. The stromata of the red blood-corpuscles contain from 10 to 12 per cent. lecithin, the digestion precipitate

[1] [*Vide* Note B, p. 300.]

certainly not under 50 per cent.; and yet all these substances form clear solutions on treatment with aqueous dilute alkali. Lecithin itself is not in the least dissolved by weak alkali.

Further, it is not at all clear that lecithin is easily 'carried down' from its solutions mechanically except when it occurs associated with proteid, and proteid is the 'carrying-down agent,' for there is not the slightest proportionality between the bulk of the precipitate and the amount of lecithin separated, but quite the contrary. The digestion precipitate from tissue-fibrinogen, though very slight in bulk, is enormously rich in lecithin. The same bulk precipitate from the original fibrinogen solution contains certainly one hundred times less.

It is clear, therefore, that, unless we suppose the lecithin to be chemically united with proteid, there are great difficulties and very much vague assumption. A-fibrinogen, the pseudo-crystalline substance, contains lecithin, certainly not under 5 per cent. (*vide* p. 283). The same general reasoning which renders it probable that A-fibrinogen is a chemical individual would be strongly in favour of the lecithin of tissue-fibrinogen being in chemical union.

VIII. *On the Processes termed 'Coagulation'*

1. *Coagulation studied on extravascular plasmata.*—In dealing with this, nothing but perfectly pure, absolutely hæmoglobin-free plasmata are used. The most scrupulous precautions are adopted as to complete removal of form-elements. In dealing with salt-plasma the greatest care must be taken to ensure that the admixture should take place with extreme rapidity—an absolute necessity, the non-recognition of which completely frustrates all attempts at exact work. Dog's blood chiefly used. For salt blood large animals with rapid flow from carotid (dogs); horses also from carotid.

I shall confine myself chiefly to two aspects of the chemical side of coagulation processes—*i.e.* 1, fibrin formation; 2, fibrin-ferment formation.

These two aspects by no means exhaust the question. There

is a third subject, in particular, to which I have already alluded—viz. the formation of new altered fibrinogens simultaneously with the fibrin. I shall, however, but briefly allude to this division, as my knowledge of it from a chemical standpoint is not yet sufficient.

2. *Coagulation of A-fibrinogen in peptone-plasma.*—As it separates from the plasma, A-fibrinogen is certainly not fibrin. Its solubilities are quite different. Nevertheless it speedily undergoes changes in the direction of conversion into fibrin. *Simple cooling*, therefore, of peptone-plasma suffices to produce a certain amount of *fibrinous coagulation.*

As I have already stated, A-fibrinogen on simple separation undergoes 'alteration,' which I regard as analogous to the 'alteration' of all the other fibrinogens (which 'alterations' are shown, as we shall shortly see, in a very high degree by the fibrinogen of plasma). But the case of A-fibrinogen is special; its separation takes place in a fluid which is not indifferent, for it can be shown distinctly that A-fibrinogen possesses in the most marked degree the power of influencing the coagulation of the plasma *in toto.*

It is impossible to collect A-fibrinogen except in an altered condition. Its appearance and physical characters are exactly like fibrin, its solubilities are in the main similar, its distinguishing features are its solubility in dilute HCl and its becoming opaque with acetic acid, together with its behaviour on digestion.

Now one obtains every gradation between typical A-fibrinogen and typical 'fibrin,' that is to say, the degree of 'alteration' shown by A-fibrinogen varies in different cases. In all cases it separates as a substance quite different from fibrin; the alteration may never go further than what I have described previously (*vide* section, 'A-fibrinogen,' p. 274), but there is every gradation between this alteration and the formation of true fibrin.

An important property of A-fibrinogen is its power of inducing coagulation in peptone-plasma when a current of CO_2 is passed through it. As I have pointed out in Section I., a commencing precipitation is helpful to this interaction. Now

A-fibrinogen can be collected and re-added to plasma, and again confer on the latter the power of coagulation with CO_2. But this is not invariably the case; if the change of the fibrinogen have gone too far in a 'fibrinous' direction, the addition is then without influence.

A-fibrinogen, then, separates out and remains for a certain time a substance quite distinct from fibrin. But it frequently passes, without any re-solution or any further treatment, into fibrin. I do not in the least mean to assert that in these changes A-fibrinogen is converted bodily into fibrin. Chemical interaction probably takes place between the substance and the body of the plasma. But if two portions of the same plasma are taken and cooled, and one portion examined quickly and the other portion after an interval, the precipitate is, in the former case, A-fibrinogen; in the latter case, either fibrin, undistinguishable from the ordinary fibrin obtained from the coagulation of plasma, or something intermediate between the two.

There are two important factors which favour the more or less fibrinous modification of A-fibrinogen. These two factors are under certain conditions helpful to one another.

Factor I.—I have alluded to what I call strong and weak peptone-plasma. Strong peptone-plasma is much less readily coagulable with CO_2 than is weak. From previous statements it will be remembered that the coagulation of peptone-plasma with CO_2 is apparently due to the interaction of A-fibrinogen with the rest of the plasma. Now the fibrinous alteration of A-fibrinogen on simple cooling invariably takes place much more rapidly, and is more complete in weak than in strong peptone-plasma.

The position is, therefore, as follows (*vide* Section I. p. 244):—

1. With CO_2 genuine undoubted coagulation of plasma.

2. Partial separation of A-fibrinogen necessary condition of CO_2 action.

3. Interaction with CO_2 much more rapid in 'weak' peptone-plasma.

4. Simple separation in weak peptone-plasma always leads to greater alteration of A-fibrinogen in the direction of fibrin.

I consider, on these grounds, that the fibrinous alteration of A-fibrinogen on cooling is due partly to an interaction between the A-fibrinogen and the rest of the plasma.

Factor II.—The second factor is very remarkable. It depends on the greater or less quantity of separated A-fibrinogen distributed in a given quantity of plasma. It is ' Massenwirkung.'

If you take two equal quantities of peptone-plasma moderately peptonised, place one (1) in a flat large dish and allow to cool slowly, the other (2) in a test-tube and allow to stand upright (both being otherwise under the same conditions), a widely different distribution of A-fibrinogen takes place in the two cases. In the one case (1) it sinks over a wide area, and is in intimate contact with a large part of plasma; in the other (2) it sinks to the bottom of the tube, and is in intimate connection with a small part of plasma.

Supposing we assume, as I have pointed out is probable, that the precipitated A-fibrinogen acts as a new fibrinogen added to the plasma, we shall in the case (1) have a greater quantity of A-fibrinogen acting on a given quantity of plasma than in case (2).

Now if reference be made to statements in the section on Quantitative Determinations (p. 293), dealing with the influence of tissue-fibrinogen on plasma, it will be seen that a certain quantity of tissue-fibrinogen must be added to produce an obvious result. A weighable quantity can be added without producing any coagulation. And in the case of A-fibrinogen, when it is allowed to collect and interact with only a limited quantity of plasma, as in the test-tube, the 'change towards fibrin' is infinitely greater and more rapid than when it is collected in flat dishes and the interaction spreads over a large area.

In dealing with the fibrinogens—*i.e.* the substances which in animal fluids are more immediately concerned in coagulation— it has been pointed out that they are by no means simple substances, that in particular they contain a very large percentage of lecithin. It has always been assumed that dealing with the material substratum of fibrin we had only to consider alterations and modifications of proteid. So far as I can judge, the view which many chemists hold on the question of coagulation is much as follows : Proteids are substances which easily pass from the state of apparent solution to the coagulated state. In the coagulation of the blood a part of the proteid, probably under a ferment influence, passes into the solid state at a lower temperature than usual. There is, therefore, nothing so especially remarkable about this. It is part of the general chemical question as to what is the nature of the change in the passage of proteid from the soluble to the insoluble modification. But this notion is utterly wrong. In the coagulation of the blood the formation of fibrin is but one thing—a striking phenomenon—but only a very small part of the whole process. Although it is paradoxical, coagulation processes occur without any coagulation. The formation of fibrin is the possible outcome of a fibrinogen interaction, but only one part of a most complex chemical interaction.

There are two processes in coagulation, I have already stated, which can be considered in something approaching chemical detail —the formation of fibrin and the formation of fibrin ferment. I pick out these two for the present, because the one is a very obvious, easily recognised process (fibrin formation) ; but it is essential to consider it with the formation of fibrin ferment, not because the fibrin ferment formation is a constant concomitant of fibrin formation, as this is not the case, but for the reason that in coagulable fluids, especially if there is any tendency to spontaneous coagulation, there are many influences, the chemical bearing of which is quite obscure, which may accelerate the formation of fibrin. Since the formation of fibrin ferment essentially depends on chemical conditions, it is of the greatest value as a control in coagulation experiments. (For special precautions needful in testing for fibrin ferment, *vide* Section I. p. 222.)

Now in the coagulation of shed blood all admit that the formation of fibrin and the formation of fibrin ferment occur. It is certain that neither of these bodies is present in blood when

U

it leaves the vessels, and yet it is certain that fibrin forms and that the serum contains fibrin ferment. Hence it has long been known that, chemically considered, the coagulation of the blood involves more than the question of proteid coagulation. But, although it has long been known, it was not till my work on the coagulation of corpuscle-free plasmata that it came fully to the front. The fibrin ferment formation was wrapped up in a morphological expression. It was stated to be formed by the ' breaking up' of white blood-corpuscles. However true this may be, it is obvious that, chemically, we get no further, and the notion of an enzyme action necessarily drew for a time one important chemical process out of the chemist's vision. Hence the misunderstandings in coagulation.

I have shown that fibrin ferment formation goes on in plasma—*i.e.* goes on *in a fluid.*[1] The chemical conditions of its investigation are therefore now possible, and its origin is not necessarily connected with form-elements. It is, in fact, no longer a morphological expression.

To illustrate this I take the following experiment :—

Peptone-plasma is quite free from fibrin ferment. A solution of tissue-fibrinogen is added, clotting ensues rapidly; the serum from this coagulation contains abundant fibrin ferment.

Now we can investigate more nearly the chemical conditions under which these processes occur. Tissue-fibrinogen is not a simple substance, it undoubtedly contains *lecithin.* (There could, I presume, be no dispute that the stromata of the red corpuscles contain *lecithin.*) And if, instead of adding tissue-fibrinogen, we add lecithin[2] prepared from tissue-fibrinogen, clotting also

---

[1] ' On the Origin of the Fibrin Ferment,' *Proc. Roy. Soc.* 1884. [Collected Papers, p. 119.]

[2] *Lecithin.*—The lecithin used for these clotting experiments I prepared with the greatest care. The solution and treatment with alcohol and ether was made over and over again. I have also treated (1) brain lecithin, and (2) lymph-gland lecithin prepared with platinum chloride as follows :—1, Solution absolute alcohol. 2, Precipitation $PtCl_4$. 3, Washing with absolute alcohol. 4, Solution in chloroform, clear filtrate obtained. 5, Re-precipitation—alcohol, and treatment of mixture with H_2S, 8–9 hours required. A little solid Na_2CO_3 was added to neutralise any acid formed. 6, Evaporation, re-solution, treatment with freshly precipitated hydrate of silver. 7, Removal of

ensues; fibrin ferment formation also takes place. This is a fact which is quite beyond any doubt;[1] and it is of fundamental importance with regard to all our ideas concerning the chemistry of proteid matter.

The extremely frequent association of lecithin with proteid has long been known. If now, in a process like coagulation of the blood, it can be shown definitely that lecithin *can* play an important chemical *rôle*, it is absurd to suppose that the presence of lecithin with proteid means nothing. At the present time it is customary to divide up proteids by their solubilities, their being precipitated by this or that, or coagulated by this or that. It is known that we are not dealing with a single proteid, but for convenience we ignore that there is another substance present, and we say the substance is a particular modification of proteid! We treat an organ or a fluid with this and that reagent. We see a coagulation occurs at such and such a temperature. Proteid, varieties of proteid! Just so we might call chlorine, hydrochloric acid, chloride of sodium, varieties of chlorine; sodium, soda, carbonate of soda, varieties of sodium; and the oxygen, the carbonic acid, &c. mere extractives! The procedure is as follows: A complex mixture is examined, its properties detailed. The mixture is then carefully boiled for hours with alcohol, &c. &c.; elementary

traces of silver by H_2S. 8, Re-solution and re-evaporation some six or eight times. The lecithins so prepared give all the reactions of lecithin, and I have most carefully tested their influence on coagulation. They possess the power in the highest degree. It must, of course, be understood that the term 'lecithin' is certainly a generic one. It is perfectly clear, from the only modern research (Hundeshagen), that, in the first place, there are polymeric substances closely allied to lecithin; and, secondly, that the chemical structure of lecithin may be a very different one in different individuals. Thus both in hens' eggs and fish eggs there occurs, apparently dissolved in fat and not connected with proteid, a very easily extractable lecithin which has no influence on coagulation, although its physical characters are very similar indeed to the lecithin obtained from proteid. It is easy to make out that there are also very decided chemical differences—*e.g.* the platinum salt of the egg-lecithin is very easily soluble, that of proteid-lecithin hardly at all. But this is a line which, while doubtless very important to investigate, is not in the scope of my inquiry.

[1] [Here, and in several other parts of this Report, the author refers forwards to sections which he intended to write.]

analysis made; conclusion, substance having the composition of the elementary analysis possesses the properties of the complex mixture.

I have chosen in this section to take ferment formation as the additional field of observation, because nothing can more clearly bring out the chemical bearing of lecithin in coagulation. Now, there is not the slightest doubt (I have given many instances of such observations) that lecithin is most effectual in causing the appearance of fibrin ferment. It is quite easy to obtain specimens of peptone-plasma in which the CO_2 coagulation and the lecithin CO_2 coagulation are not very strikingly different until we come to estimate the ferment, and then we find that in the coagulation in presence of lecithin the ferment formation is increased a hundredfold. There is no other proximate principle which I have found to possess this power in any way. Great production of ferment is evoked by a variety of complex substances (*e.g.* all the 'fibrinogens'), but they all contain lecithin in abundance, and from consideration of this fact I have never had the slightest doubt of the chemical influence of lecithin.

It must be understood that although lecithin causes clotting and the appearance of fibrin ferment, the clotting is not caused by the fibrin ferment. They are generally associated phenomena, but that there are not two processes—*e.g.* 1, formation of ferment influenced by lecithin; 2, formation of fibrin influenced by ferment—is seen from the following :—

1. Lecithin produces clotting under conditions in which the strongest fibrin ferment is ineffectual.[1]

2. Under special experimental conditions lecithin produces coagulation, but *no ferment*. (Details and analysis of process to follow.)

Not only does lecithin unquestionably enormously favour the formation of fibrin ferment, *but there is probably never any ferment formed unless either free lecithin, or lecithin in a particular form (loosely combined with proteid) be present.* This last statement is at present only an extreme probability. It is not a proven fact, because we are so utterly ignorant of the possible

---

[1] Vide 'Uebersicht ü. d. Gerinnung.' [Collected Papers, p. 186.]

relationships of proteid and lecithin, and it is so very difficult to get at the problem in a manner satisfactory to chemical criticism. I will merely shortly illustrate the point.

A-fibrinogen, under given conditions which have already been mentioned, produces coagulation and ferment formation. After precipitation, it may again be added to plasma and still do this. But it does not do this if it has undergone too much 'alteration' (see page 274).

Now A-fibrinogen, both before and after this alteration, contains lecithin, and it is impossible to say whether more or less; but coincident with its 'alteration'and loss of *coagulation-producing* power, it undergoes remarkable changes in its chemical behaviour. The alteration consists in the fact that it has become fibrinous instead of merely fibrinogenic; this is an obvious alteration. But the chemical point is that its behaviour on digestion is profoundly modified. As I have pointed out, the fibrinogens have a very characteristic behaviour on artificial digestion. They readily yield a precipitate which is extraordinarily rich in lecithin. But after they have undergone the fibrinous modification they do not do this, and consequently I am inclined to think that in the latter case—*i.e.* in clotting—the lecithin-proteid relationship must have been altered.

The exact meaning to be attached to the experiment is open to question, and requires very full treatment, which I am not at present prepared to give.

This is the only illustration I will give at present. It is a very wide and important question, which I am and have for a long time been dealing with.

Although I have confined myself to the occurrence of clotting and the formation of fibrin ferment, it must be thoroughly understood that these are but a limited part of the chemical aspect of coagulation.

IX. *Quantitative Relationships in Fibrinogen Interactions*

From the general account given in my Croonian Lecture, it was clear that the process described under the head of Fibrinogen

Interaction was quite different from any fermentative process. This will be the more apparent from a consideration of the following quantitative determinations. The quantitative estimations adduced are purposely limited in number, because, as will be clear from subsequent statements, the greatest caution is necessary in drawing conclusions from such estimations.

The previous quantitative work seems to have had in the main but one object, viz. to determine whether or not there was any 'fibrinoplastic' influence to be admitted in coagulation. Schmidt, in fact, relied mainly for the support of his theory on certain quantitative determinations. Certain transudation fluids yielded unmistakably larger quantities of fibrin when, in addition to ferment, paraglobulin was added. This observation has been adduced to prove that paraglobulin takes a material part in coagulation, and it is really the only point on which Schmidt relied. But, as Hammarsten has pointed out, it is not necessarily a proof that paraglobulin is at all concerned in coagulation. It is easy to adduce an instance parallel to those brought forward by Hammarsten from my own experience.

Ferment added to peptone-plasma causes little or no coagulation, but with the addition of a stream of CO_2 there is infinitely more coagulation. It does not, however, follow that the CO_2 is a factor in the material building up of fibrin. The same objections can be raised, and have been raised, to Schmidt's results. Now, although I am quite certain that Schmidt afforded no convincing proof of a 'fibrinoplastic action' whatever, it appears to me that Hammarsten was wrong in his endeavour to show that there was no 'fibrinoplastic action' in Schmidt's sense. I think Schmidt undoubtedly observed one special instance of a very important process which Hammarsten entirely overlooked, and which I have termed the 'fibrinogen interaction.'

The attempt to prove the participation of 'fibrinoplastin' by quantitative determination of the amount of fibrin formed[from any particular coagulable fluid only narrowed extremely the full consideration of the process of coagulation. It could only be effectual in the case where the quantity of fibrin formed exceeded the total fibrinogen of the coagulable fluid. This condition has, however, never been shown to occur, and hence Hammarsten's objection that the ' fibrinoplastic action ' was merely an indirect influence must always hold good.

Existence of the ' fibrinogen interaction' could not possibly be adduced entirely from quantitative estimations of the fibrin formed ; for, as I have pointed out, a very pronounced fibrin-ogen interaction may take place without any fibrin at all being formed, but the very opposite condition is set up—*i.e.* in intra-vascular injection of small quantities of fibrinogen the *negative phase* is evoked.

For the material participation of tissue-fibrinogens (serum-fibrinogen, &c.) in the process of coagulation, therefore, it is obvious that only secondary importance can be attached to quantitative fibrin determinations, although they are not without interest and importance.

The following observations on this subject refer to the inter-action of peptone-plasma and a solution of fibrinogen from the thymus gland. I would again recur to the point that this inter-action is not necessarily identical with processes occurring in nature.

On adding a solution of tissue-fibrinogen to peptone-plasma coagulation occurs and the tissue-fibrinogen disappears—*i.e.* the resulting serum does not contain tissue-fibrinogen. This can be very clearly demonstrated when using the fibrinogen from thymus, because, as I have pointed out, this substance is extremely insoluble in acetic acid, whereas the fibrinogens of plasma and of serum are easily soluble in excess.

But the power of a given quantity of plasma to effect this change is strictly limited, as will be obvious from the following :—

Exp. A.—By tests the quantity of tissue-fibrinogen which could be added to a specimen of plasma and certainly disappear was deter-mined, when it was found that 20 c.c. of a solution of tissue-fibrinogen (from thymus) could be added to 30 c.c. of plasma and disappear. If more were added the absence of tissue-fibrinogen from the serum was doubtful. The acetic acid test was used. The 20 c.c. of tissue-fibrinogen solution contained ·120 grm. tissue-fibrinogen.

Exp. B.—25 c.c. plasma totally clotted (*i.e.* the serum gave no further trace of clotting on adding tissue-fibrinogen) with 10 c.c. tissue-fibrinogen ; clotting occurred in five minutes. Weight of fibrin, ·089. The dry weight of tissue-fibrinogen was ·056. The plasma contained ·548 per cent. fibrinogen. (For method of determination see Note A at the end of this Report.)

In this experiment the weight of the fibrin is greater than that of the tissue-fibrinogen added. Some of the fibrin must, therefore, have come from the plasma.

The total fibrinogens concerned in the reaction were :

Of plasma	·137
Of tissue-fibrinogen . . .	·056
	·293
Of fibrin	·089

The amount of fibrin is thus much less than that of the fibrinogen of the plasma alone.

Exp. C.—20 c.c. plasma treated with CO_2, gave slow clotting. Fibrin formed, 0·71.

20 c.c. plasma with tissue-fibrinogen in excess. Fibrin formed, 0·82.

20 c.c. of the plasma contained 1·28 fibrinogen.

In this experiment there is a slight increase of fibrin in the tissue-fibrinogen coagulation. But determinations of this nature give us no information as to the tissue-fibrinogen being materially concerned in coagulation, and the amount of CO_2 coagulation varies greatly with external circumstances. It may be nothing or very extensive. Experiments of this kind are analogous to the fibrinoplastic determinations of Schmidt. They really only obscure the coagulation question.

Proportionality between the amount of tissue-fibrinogen added and the fibrin formed. (The proportionality is marked, as explained in Section I., only when using strongly peptonised plasma.)

Exp. D.—(1) 20 c.c. plasma ; 5 c.c. solution tissue-fibrinogen. Fibrin, ·025 grm.

(2) 20 c.c. plasma, 10 c.c. solution tissue-fibrinogen. Fibrin, ·061 grm.

Clotting in (1) did not occur for twenty minutes and was very incomplete. In (2) it took place in five minutes.

This difference in time in the action of large and small quantities of the fibrinogen is always marked. The serum of experiment (1) remained quite clear for hours; clotted, however, through and through on addition of more tissue-fibrinogen.

In Experiment B the amount of fibrin formed is greater than that of the tissue-fibrinogen added. It may be less, for small but weighable quantities of tissue-fibrinogen may be added to peptone-plasma without causing any coagulation at all.

In Experiment B 10 c.c. tissue-fibrinogen ('056 dry weight) acting on 25 c.c. plasma produced total coagulation yielding 0·89 fibrin. With the same plasma, 2 c.c. tissue-fibrinogen ('01 grm. solid) acting on 15 c.c. plasma produced none.

It will be obvious, from the numbers quoted above, that estimations of the amount of fibrin formed cannot give us precise information as to the chemical process which has taken place, and for this reason. The fibrin formed is always less than the fibrinogen present; in other words, a part of the fibrinogen does not separate as fibrin, but remains in solution. Now quantitative estimations cannot give us full information unless we can exactly characterise and estimate the total quantity of the products of fibrinogen interaction—i.e. not only the fibrin but the soluble substance or substances--and this we cannot at present do.

The soluble products of a fibrinogen interaction not only vary greatly in quantity but they vary in quality, and it is not possible to give any positive method of determination at the present time. The direction in which my researches have gone will be later pointed out. The study of the process of coagulation in peptone-plasma, and of the intravascular effect of fibrinogen injections, forces this question of the soluble products of 'coagulation processes' very much to the front. Probably it forms a part of what I have called the negative phase of the fibrinogen interaction, but not the whole.

In dealing, then, with the question of whether a particular proteid enters materially into the formation of fibrin, it must be clearly understood that the weight of fibrin formed does not give us information.

To take Experiment C. With CO_2, 20 c.c. of the plasma gives ·071 fibrin. The total fibrinogen in 20 c.c. of the plasma is ·128; a great part, therefore, of the fibrinogen of the plasma is not separated as fibrin.

The same quantity of the plasma is treated with tissue-fibrinogen, and yields ·08 fibrin—*i.e.* a slight increase;[1] but could the soluble products of the interaction be estimated, they would probably be greater too, and until this can be accurately done the quantitative estimation of fibrin is of little avail. Evidence obtained in other ways, to which I shall subsequently refer, is of far greater force than quantitative determinations.

NOTE A

FIBRINOGEN DETERMINATION

In order to avoid side issues, and to make it clearly understood how little value is really to be found in quantitative estimation, I have not discussed the method of fibrinogen determination. This is naturally a matter of great importance, but at present we are not able to do much more than guess at the facts.

The methods which have been adopted by previous investigators have been those of taking the fibrin formed as a measure of the amount of fibrinogen present, or of taking the amount of coagulum formed on heating to 56° as the measure of the fibrinogen. Now it is possible that under certain conditions the amount so obtained might give results mainly corresponding to the amount of fibrinogen. But it has also been shown that under certain conditions—*i.e.* when ferment acts on certain solutions of fibrinogen—a part of the fibrinogen always remains in solution, and also that when solutions of fibrinogen are boiled a part remains in solution, and that the ratio of soluble to insoluble parts is variable (Hammarsten). Clearly, then, neither of these methods can pretend to accuracy, and they are in most cases instances of working entirely in the dark.

The method I have used is open to objection, but it is at the present time the only one available. It depends on the following facts. Paraglobulin is soluble, though not easily, in saturated solution of NaCl. Fibrinogen is not thus soluble. I therefore precipitate the plasma by saturation with NaCl. The precipitate is washed with saturated NaCl solution until all proteid reaction is gone from the wash-fluid ; this ought to remove the paraglobulin. The precipitate is then dried at 100° C. and washed with water till freed of all salt. The results are probably almost certainly too low.

[1] The slight increase, ·07–·08, is not an indication of ' fibrinoplastic action, because the CO_2 coagulation is most variable.

I give an illustration of the importance of this question of fibrinogen determination.

As is well known, Hammarsten states that the serum contains more paraglobulin than the plasma, and this result has very generally been adopted as true. I quote an illustration of the sort of evidence on which it rests. His method is this : he takes the cooled filtered plasma and the serum of the same animal. In both he determines the total globulins by the $MgSO_4$ method. In the plasma he determines the fibrinogen and subtracts this from the total globulins, the residue representing the paraglobulin.

It follows from this method that if the serum really contains more paraglobulin than the plasma, it must come from something outside the plasma. The source is supposed to be the white corpuscles—this being a pure hypothesis.

The following is an instance :—

Globulins in plasma	4·87
Globulins in serum	4·48
Fibrinogen of plasma (a) determined by heating .	0·32
„ „ (b) determined by fibrin . .	0·62
Paraglobulin of plasma in case (a)	4·55
(i.e. more than serum.)	
Paraglobulin of plasma in case (b) . . .	4·25
(i.e. less than serum.)	

Now, although the percentage error of determination in the two cases is not very great, it is obviously of great importance in dealing with such small quantities to know whether 0·32 is the amount, or the double of that, 0·62 ; because in this particular instance it is just this difference which tells for or against the assertion of Hammarsten.

There is also to be considered in regard to this result that neither number, by Hammarsten's own showing in subsequent papers, represents the real amount of fibrinogen, but only an unknown fraction, and also that the filtered cold plasma has already lost a considerable and unknown quantity of its fibrinogen (vide Section I.). All these numbers of Hammarsten are, therefore, purely fancy numbers, and it is impossible to deduce any result whatever from them.

Now Hammarsten is an observer of whom I have the highest opinion, and I introduce this criticism merely with the object of showing that the attempt to solve the coagulation question by such chemical methods is at present likely to lead to false conclusions, and by giving such false conclusions an appearance of scientific accuracy to tend to perpetuate error

It cannot be too clearly understood that we must have the chemical problems of coagulation sharply marked out ; we must know for certain what our results mean. To continually draw conclusions which in reality rest on a pure assumption is a futile proceeding, because the assumption may turn out to be wrong.

NOTE B

ON THE REACTION OF FIBRINOGENS TO GASTRIC DIGESTION

It is this reaction with an artificial digestive fluid which is so characteristic of the fibrinogens, and which renders them so closely allied to the casein of milk.

I think these bodies have long been known and spoken of as nucleo-albumins. It is an unfortunate expression. It seems to have arisen in the case of Plosz, working with the fibrinogen of the liver. On artificial digestion this yielded an albuminous precipitate, soluble in alkali, and very rich in phosphorus. Plosz concluded this was nuclein, without investigating the condition of the phosphorus.

Now there appear to be proteid substances containing large quantities of phosphorus not in the form of lecithin, at any rate not extractable by alcohol. The term nuclein was certainly not meant to include proteids, which were mixtures or possible compounds of proteid and lecithin. And these digestion precipitates of the fibrinogens contain very little, if any, phosphorus after complete extraction with alcohol ; and, owing to the great difficulty of completely removing the lecithin, it is difficult to say whether the phosphorus which remains behind in small quantities is due to incomplete extraction or not.

I am at present occupied with determining, as accurately as possible, the quantitative relationships of the lecithin to the proteid in the precipitate. The following numbers are approximations ; they certainly underestimate the quantity of lecithin :—

A. Fibrinogen from Thymus.
On digestion with artificial gastric juice gave the usual precipitate.
    Weight of dry washed precipitate .   .   . 1·68 gram.
    Weight of lecithin extracted from precipitate ·24 ,,
    Percentage amount of lecithin in precipitate . 14·2 %

B. Fibrinogen from Plasma.
    Dry weight of digestion precipitate . . 1·12 gram.
    Weight of lecithin extracted .   .   . ·32 ,,
    Percentage amount of lecithin . . , 28·5 %

APPENDIX TO PART

APPENDIX TO PART I

NOTE ON THE COAGULATION OF THE BLOOD [1]

IN a paper read before the Royal Society, April 26, 1888, Dr. Halliburton offers some criticism of my views respecting the coagulation of the blood. In this note I shall briefly summarise and traverse the objections Dr. Halliburton raises to my theory and experiments.

I. Dr. Halliburton suggests that the substance I call 'A-fibrinogen'—which I obtained by cooling peptone-plasma—is not a normal constituent of the blood-plasma, but that it is a precipitate of a hemi-albumose, supposed by him to be present in the peptone which is injected into an animal for the purpose of obtaining peptone-plasma. I do not use Witte's peptone, as Dr. Halliburton appears to have done, on account of its recognised impurity, but that obtained from Dr. Grubler's well-known laboratory in Leipzig. This peptone is prepared according to Henniger's method. A 10 per cent. solution of it in ½ per cent. solution of sodium chloride is quite clear after filtration.

It gives no precipitate on cooling to zero.

It disappears wholly from the blood within one or two minutes after injection.

Finally, A-fibrinogen has properties absolutely different from the peptone injected.

Dr. Halliburton appears to think that this substance, A-fibrinogen, exists only in peptone-plasma.

I stated in a paper read before the Royal Society in 1885, 'On a New Constituent,' &c., that it was also present in salt-plasma, and I gave details concerning it in the Croonian MS., which is in the archives of the Royal Society. I explained at length in the paper referred to by Dr. Halliburton, and published in Ludwig's

[1] [From the *Proc. Roy. Soc.* vol. xliv. 1888.]

'Festschrift,' 1887, why there are, as has long been known, two varieties of salt-plasma—namely, one containing, as I showed, no A-fibrinogen, this being not spontaneously coagulable ; the other containing it, and therefore being spontaneously coagulable.

II. Dr. Halliburton further asserts that whereas in the abstract of the Croonian Lecture I described a body, B-fibrinogen, in the paper in Ludwig's 'Festschrift,' published shortly afterwards, this body was not mentioned, or had become identical with the fibrinogen of Hammarsten. This statement is totally incorrect, for on page 228 of Ludwig's 'Festschrift' there will be found a paragraph headed 'B-Fibrinogen,' and on the following page this passage occurs : 'Man sieht also, dass das Fibrinogen von Hammarsten in Plasma einen Vorgänger hat, welcher andere Eigenschaften besitzt, und ich bezeichne diese Substanz als "B-Fibrinogen."' The differences between the two bodies here referred to are precisely those mentioned in the abstract of the Croonian Lecture, and are shortly as follows :—

(a) B-fibrinogen does not clot with fibrin ferment, but it does clot with leucocytes and other animal and vegetable cells.

(b) It clots with substances which can be obtained from these animal and vegetable cells in large quantities by extraction with water. These substances I call 'tissue-fibrinogens.'

(c) It further clots with lecithin.

Hammarsten's fibrinogen, in remarkable contrast with the properties of this body, does not clot with leucocytes or other animal or vegetable cells, nor does it clot with substances called 'tissue-fibrinogens' nor with lecithin.

I would here add that the fibrinogen in most transudation fluids is similar to Hammarsten's fibrinogen. I have clearly indicated these differences in previous publications.

III. With regard to Dr. Halliburton's remark on the relation of lecithin to clotting, I may say that it not only gives rise to clotting in peptone-plasma and cooled plasma, but in a solution of fibrinogen isolated from salt-plasma and in the plasma obtained from the blood after the injection of tissue-fibrinogen. In discussing the experiments on the behaviour of cooled blood towards lecithin, Dr. Halliburton does not recount the details of the experiments, and hence he conveys a misleading impression of the same. It is necessary for these experiments to use a finely particulate and yet thick emulsion of lecithin, for the following very obvious reasons. The lecithin is insoluble in the salt solution into which the blood is received, and a large quantity of blood being received into a relatively small quantity of the salt solution, the lecithin does not come into contact with all the plasma unless a fine thick emulsion be used.

The fact that fluids free from lecithin produce clotting, in no way disproves the contention that lecithin is an essential factor in coagulation, since every variety of fibrinogen contains lecithin. Lecithin is, next to proteid, the most widely distributed substance in the animal organism. As Hammarsten has well said, 'it has been found wherever it has been looked for.' Whenever I have stated that lecithin is present in any fibrinogen, I have prepared it and tested for it in the way I have previously repeatedly described in the papers Dr. Halliburton quotes.

IV. The criticisms which Dr. Halliburton passes upon my discovery that tissue-fibrinogens cause intravascular clotting when injected into the living circulation, can hardly be regarded seriously ; for he asserts that the tissue-fibrinogen is a slimy mass, and causes clotting by mechanically plugging the vessels, whereas if he had repeated my experiments he would have found (1) that the fibrinogen is not at all slimy, and (2) that it can hardly be supposed to cause clotting mechanically, since it passes through the right heart, then the capillaries of the lungs, next the left heart and aorta, and finally the capillaries of the alimentary canal, before it first causes clotting, i.e. in the portal vein in the dog.

ON THE QUESTION OF COAGULATION [1]

Dr. KRÜGER has lately published in this journal (1887, Part 2) a paper consisting mainly in a criticism of my 'Theory of Coagulation,' [2] which has recently appeared. In it the following sentence occurs : 'Wooldridge is certainly not in error in maintaining that lecithin plays a part in coagulation, since it is a fact already definitely ascertained by J. v. Samson-Himmelstjerna and A. Nauck that not only lecithin but a whole series of substances obtained in the disintegration of proteids are concerned in fibrinous coagulation ; only, however, as sources of ferment, i.e. as substances from which the fibrin ferment is readily split off under appropriate conditions. These are described more precisely by the authors already quoted in papers of which Wooldridge has taken no notice whatever.'

The accounts of Samson-Himmelstjerna and Nauck appeared in 1885 and 1886 respectively. In my 'Theory of Coagulation' I

---

[1] [Translated from *Zeitschrift für Biologie*, 1888, p. 562.]

[2] Ludwig's *Festschrift*, 1887. [Collected Papers, p. 186.]

have endeavoured to give a short exposition of the work that I had already communicated.

In 1883 I discovered the close connection existing between lecithin and the formation of fibrin, and in the same year I published three papers on this subject.[1] I also considered the relation of lecithin to the formation of fibrin ferment in the 'Proceedings of the Royal Society' for 1884.

As the papers of Samson-Himmelstjerna and Nauck appeared two or three years later, it was obviously impossible for me to quote them. Should Dr. Krüger still maintain that these facts have been firmly established by these authors, I can only say that I am very pleased to have my discovery thus confirmed. With regard to the other 'retrogressive proteid metamorphoses' and their relation to fibrin formation, I will gladly leave the priority of the discovery to these gentlemen, since I am of opinion that they are of no importance at all in the coagulation of the blood.

Dr. Krüger has somewhat misconceived my position with reference to the co-operation of cellular elements. I certainly do not deny that leucocytes from lymph-glands and other tissue-cells have eminently the power of inducing coagulation in extravascular plasma ; in fact, I was the first to discover this.[2]

But they lose this power so soon as they are introduced into the circulating blood. In my paper in Ludwig's 'Festschrift' I have shown that if leucocytes from lymph-glands be injected into the circulation they do not produce any intravascular clotting, and, moreover, the blood when drawn off will not coagulate ; they must, therefore, have lost their power of causing clotting during their sojourn in the living blood. It follows that the effects observed on adding cells to plasma cannot be taken to prove the participation of white blood-corpuscles in coagulation.

The further objections of Dr. Krüger will be considered in a detailed account of my work on coagulation.

THE COAGULATION QUESTION[3]

In a paper read before the Royal Society in April, 1888, and recently in this journal, Dr. Halliburton has published his belief

[1] 'Zur Gerinnung der Blutes,' *Du Bois' Archir*, 1883; 'Further Observations on the Coagulation of the Blood,' *Journ. of Physiol.*; 'On the Coagulation of the Blood,' *Ibid.* [Collected Papers, pp. 106, 100, 113.]

[2] *Du Bois' Archir*, and *Proc. Roy. Soc.* 1881. [Collected Papers, p. 89.]

[3] [From the *Journ. of Physiol.* 1889. (This paper was preceded by the follow-

that the fibrin ferment is a cell globulin, and also a somewhat extensive criticism of my work on coagulation.

In the following discussion of the chief points which Dr. Halliburton has raised against my work and views I will first take the last point handled by Dr. Halliburton in his 'criticism.'

An important feature in my work on coagulation is the production of intravascular clotting by the substances I call 'tissue-fibrinogens.'

As an explanation of this intravascular clotting, Dr. Halliburton said, in his Royal Society paper, there was fair reason to suppose the solutions of these substances were of a very slimy nature, so that they would, when injected, naturally block up the small vessels and thus cause coagulation ; just as an iron wire will cause coagulation when introduced into the sac of an aneurism.

From the very first papers I had published on this subject (1886) it was clearly impossible that this suggestion could be correct, and in the light of what I had subsequently published, previous to Dr. Halliburton's first paper, it was manifestly out of the question. With the greatest ease Dr. Halliburton could, by repeating my experiments, have seen whether his assertions were true or not. This he has now done, and he withdraws his observations unreservedly.

In his paper before the Royal Society Dr. Halliburton stated also that I had spoken of a substance, B-fibrinogen, in my abstract of the Croonian Lecture, and that this substance had disappeared from the paper which was published in Ludwig's 'Festschrift,' or had become identical with Hammarsten's fibrinogen. Now since a large part of my paper in Ludwig's 'Festschrift' was devoted to describing B-fibrinogen, and the differences between it and Hammarsten's fibrinogen, Dr. Halliburton's assertions were quite unintelligible to me.

In my researches on coagulation I lay great stress on a substance which separates from peptone-blood-plasma on cooling ; I call it 'A-fibrinogen.' I have taken the greatest and most conscientious trouble to investigate this substance, and I regard it, and I am sure with truth, as of the greatest importance in coagulation processes. Now Dr. Halliburton plainly indicated in his first paper that this

ing Editorial Notice : The terribly sad event, the untimely death of the author on June 6, by which Physiology has been robbed of a bright and enthusiastic inquirer, has laid on me the duty of carrying out by myself the final revise of this paper. It has already been revised by the author ; and I have endeavoured to make as few changes as possible.—M. FOSTER.)]

precipitate was merely an albumose mixed with the peptone injected, and was in no way a constituent of the blood. The principal point in his criticism was, and is still, that solutions of Witte's and Grubler's peptone yield on cooling a precipitate of rounded granules. I pointed out that the peptone I use (Grubler's) gives no precipitate on cooling, and it disappears entirely from the blood. I was, of course, aware at the time when I first found the 'cold body' (A-fibrinogen) that peptone (using the word as a generic term) had been obtained in rounded granules (Schmidt-Mülheim), and also that other proteids—i.e. vegetable albumen and paraglobulin—could be obtained in rounded granules ; for I had worked for a long time at Leipzig at crystallised proteids, the first step towards crystallisation always being the production of rounded granules. The mere fact that A-fibrinogen can be obtained in rounded granules is, of course, no more entirely characteristic of it than would the statement that a body was crystallisable suffice to distinguish it. But from the very first there could be no doubt that this substance was not peptone. I had seen it for a long time before recognising what it meant. As I saw it, its appearance was precisely similar to a fibrinous clot, and for a long time I thought it was a partial coagulation of the plasma due to the presence of leucocytes. I was struck by the fact, however, that this clot only appeared when I kept my plasma overnight in ice, and that it was absent from my plasma when kept in the warm room. Its characters, both chemical and physical, and its influence on coagulation, I have described in detail in my papers ' Ueber einen neuen Stoff,' &c. ; [1] ' On a New Constituent of the Plasma ' ; [2] in Ludwig's ' Festschrift,' and ' Beiträge zur Frage der Gerinnung.' [3] If Dr. Halliburton had suggested that A-fibrinogen was merely fibrin it would have been intelligible but not correct. But, although it is difficult from Dr. Halliburton's statements [4] that cooling peptonised serum and other solutions of peptone give a precipitate exactly similar to that of A-fibrinogen (Wooldridge), and that, as Neumeister has shown, the albumose or peptone injected ' exists as such or in a loosely combined condition in the blood for a time,' to conceive that he meant or means anything else than that my A-fibrinogen is but such precipitated albumose, we are now to understand (p. 280, the same on which the above statements are made) he means nothing of the sort. All he means is that albumose alters the normal proteids of the plasma.

[1] *Du Bois' Archiv*, 1884.   [2] *Proc. Roy. Soc.* 1885. [Collected Papers, p. 124.]
[3] *Du Bois' Archiv*, 1888.   [Collected Papers, p. 253.]
[4] *Proc. Roy. Soc.* vol. xliv. p. 266 ; and *Journ. of Physiol.* p. 280.

This is quite a different view, and is, of course, a necessary conclusion from my own experiments and publications. In fact, Dr. Halliburton tries to discredit my observations on this subject by—

I. Hinting that it is nothing but albumose precipitated by cold ; indeed stating, page 278, § 2, ' that the occurrence of the so-called A-fibrinogen in peptone-plasma is produced by the peptone, is supported by the fact that if one takes a solution of Witte's peptone and cools it to 0° C. a precipitate is produced consisting of rounded granules having very much the appearance of blood tablets.'

II. Asserting that, nevertheless, it is not albumose, is not the actual precipitate from the peptone injected.

The flagrant contradiction of these two ways of looking at the subject is only consummated by Dr. Halliburton's suddenly dropping the subject he raises—viz. what is the nature of the body I call A-fibrinogen ? But he proceeds to add remarks the justification for which rests on his ignorance of my researches on this subject.

For instance, Dr. Halliburton complains that in my short answer to him in the ' Proceedings of the Royal Society ' I do not give a list of the properties of A-fibrinogen. Am I compelled to rewrite all my descriptions and facts because Dr. Halliburton will not take the trouble to read them ? On page 278 of the Journal paper Dr. Halliburton says the following in reference to the occurrence of A-fibrinogen in salted plasma :—' The statement that sodium chloride plasma contains A-fibrinogen is thus a matter of pure inference ; and the reasoning adopted by Wooldridge is a very obvious case of reasoning in a circle as follows : this form of salted plasma coagulates spontaneously, therefore it must contain fibrinogen A, and because this plasma contains fibrinogen A therefore it coagulates spontaneously.'

What I have really said on this subject is as follows : Peptone-plasma free from ferment and corpuscles—

1. Yields on cooling a substance which can be collected and examined, A-fibrinogen.

2. Clots spontaneously—i.e. by means not [1] fibrin factors.

3. When it clots spontaneously yields fibrin ferment.

4. Cooled, and with A-fibrinogen thus removed, loses the power of spontaneous coagulation and of forming fibrin ferment.

5. Treated with sodium chloride solution so that the mixture contains 4 to 5 per cent. of the salt, yields no precipitate on cooling and retains the power of spontaneous coagulation.

6. If treated with $MgSO_4$, as I have described, a precipitate

[1] ? Independently of.—ED.

is formed in the plasma closely resembling the cold precipitate, A-fibrinogen.

7. After the removal of this precipitate the plasma no longer coagulates spontaneously and forms no fibrin ferment.

To judge of the action of salts on plasma we must first have it. Normal blood of the dog (*i.e.* blood the moment it leaves the vessels), treated severally with 10 per cent. NaCl solution and MgSO₄ solution, yields plasmata exactly corresponding, as regards spontaneous coagulability and *formation* of fibrin ferment, to those obtained by the action of salts on peptone-plasma.

Consequently I consider I am justified in the deduction that A-fibrinogen is present in salted plasma.

In his own paper Dr. Halliburton says : 'Magnesium sulphate plasma is generally the slowest to coagulate when simply diluted with water ; this is, *no doubt*, because some of the fibrinogen of the plasma is precipitated by the use of this salt.' It is not clear why Dr. Halliburton says 'no doubt.' It is precisely the statement I have made in consequence of my study of the action of magnesium sulphate on peptone-plasma. Nor is it clear why, with his views, Dr. Halliburton thinks the removal of this fibrinogen should delay the occurrence of clotting. Has he any evidence to show that a slight diminution in the quantity of fibrinogen makes the rest clot slower with ferment ? Why should it ?

Now there is this difference between the two kinds of salted plasma in the dog. The sodium chloride plasma clots in from three to ten minutes on simple dilution. The magnesium sulphate plasma does not clot for forty-eight hours. Such a striking difference is only explicable by the assumption that the fibrinogen removed by the MgSO₄ has a special power of initiating coagulation.

What I have said is that NaCl plasma and uncooled peptone-plasma contain a special fibrinogen (A-fibrinogen) which has the power of initiating coagulation.

Dr. Halliburton first tries to throw ridicule on my results, and then for his own purposes practically admits they are true.

He states in his paper in the Journal that *I infer that solutions of tissue-fibrinogen cause clotting in extravascular plasma,* from its action on intravascular blood. But this again is quite inaccurate.

In my paper in Ludwig's 'Festschrift,' and in a paper published in March, 1888, in Du Bois' ' Archiv,' I state and describe fully the action of tissue-fibrinogen on extravascular plasma.

I quote from Ludwig's ' Festschrift ' (which Dr. Halliburton frequently refers to). 'Diese Stoffe (Gewebes-Fibrinogen) bewirken nicht nur Gerinnung im extravascularen Plasma, wie Pepton-Plasma, sondern sie bringen auch beim Einspritzen u.s.w. Die genannten Stoffe sind, ähnlich wie Lymphkörperchen und *Lecithin*, gänzlich ohne Wirkung auf stark verdünntes Bittersalzplasma—bringen sie dagegen extravasculares Plasma zur Gerinnung, so entsteht gleichzeitig viel Ferment.'

I think the illustrations I have given above indicate fairly the nature of Dr. Halliburton's criticism.

Having sufficiently dealt with what I must characterise as a travesty of my researches, I now pass to a consideration of the actual observations of Dr. Halliburton. Dr. Halliburton, to his own satisfaction, conclusively overturns all my work by the following highly remarkable dictum : ' Now Dr. Wooldridge admits too much ; he admits that the fibrin ferment causes coagulation,' and ' The formation of fibrin either is or is not a ferment action, it cannot be both.'

Sugar is formed from starch by ferment action ; does Dr. Halliburton think it can be formed in no other way ? In such a state of mind does Dr. Halliburton approach the coagulation question. Truly, I admit that fibrin ferment sometimes causes the coagulation of fibrinogen ; to state that fibrin cannot be formed in any other way is to beg the question at issue, and consequently to assume most unwarrantably perfect knowledge of the whole process of coagulation.

The destruction my observations undergo in Dr. Halliburton's hands, when he treats them experimentally, is quite on a par with that treated logically.

He completely 'upsets one of the very foundations on which' my theory is constructed.

He describes six experiments in which he prepares, according to Schmidt's method, solutions of fibrin ferment from lymphatic glands. He consequently ' feels justified in concluding that the cells of lymph-glands yield as one of their disintegration products a substance which has the same properties as fibrin ferment.' ' This is a fact which has been observed by others ; it is, however, denied by Wooldridge.' I have, of course, never denied in the least that leucocytes might as the result of disintegration yield fibrin ferment ; quite the contrary. The real statement I have made is reproduced in the very same sentence by Dr. Halliburton himself, and among the 'others' whom he quotes as supporting his view, and diametrically opposed to me, is Rauschenbach.

Dr. Halliburton's own work on leucocytes is in many respects an incomplete repetition of Rauschenbach's.

Rauschenbach, a pupil of Alexander Schmidt, repeated and extended my original observations, using cooled plasma. He is the originator of the statement that the leucocytes from lymph-glands contain no fibrin ferment and that they have no action on certain fibrinogen solutions. He is the originator of the statement that there are 'mother substances' of the fibrin ferment. From my own papers eight years ago and from Rauschenbach's onwards there has been a complete revolution in the coagulation question.

I quote from Rauschenbach.[1]

'When the cells are treated with distilled water and immediately filtered, ten parts of the filtrate with one part of salt-plasma (the test fluid) gave signs of a weak fermentative activity (clotting in four or five hours)' (p. 27).

'The slimy mass of leucocytes which had been produced by the action of salts, treated with alcohol yielded only traces of fibrin ferment' (p. 25).

'After the cells had been acted on by distilled water for many hours an active extract was obtained ; heating to 50° quickened the formation of this active extract' (p. 27).

'Whether the traces of fibrin ferment which even the freshest collection of cells contains, and which pass into the watery extract of the alcohol coagulum and into the rapidly filtered watery extract of the cells themselves, are of vital origin or are formed during the pressing out and filtration of the mass, I must leave for the present undecided' (p. 28).

'I must expressly emphasise here that fresh pressed-out cell fluid, in a quite undiluted condition, exerted no perceptible action whatever on my fibrinogen fluids nearly free from paraglobulin. The same is true of the watery extract obtained immediately after pressing out the cells' (p. 30).

My observations in regard to the absence of ferment from fresh isolated leucocytes are quite in accord with those of Rauschenbach. As Dr. Halliburton quotes him, one would gather that this observer's statements were quite in harmony with those of Dr. Halliburton and opposed to mine ; whereas the exact contrary is the case.

Now these observations of Rauschenbach clearly point to the fact that fresh isolated leucocytes contain none or dubious traces of fibrin ferment ; that by artificially breaking them up fibrin ferment may be formed. Dr. Halliburton, in dealing with me, says that

[1] 'Protoplasma und Blutplasma,' Dorpat, 1882.

all my complicated and in his eyes apparently ridiculous observations fall to the ground because the leucocytes contain such a powerful fibrin ferment. He says, speaking of peptone-plasma : 'Leucocytes contain a very powerful ferment, therefore they cause clotting in peptone-plasma, whereas serum and Schmidt's ferment are not so powerful.'

There is not a particle of evidence in the whole of Dr. Halliburton's paper in support of this statement—the one experimental point he endeavours to make against me.

From the six experiments I have alluded to it is impossible to judge whether there is any fibrin ferment present at all. Since the only approximately reliable experiment, with strong magnesium sulphate plasma, is spoilt and rendered dubious by the very slight dilution of the plasma, and as no weight or measures or comparisons with serum are given, there is not even an attempt to prove his statement.

Lymph-cells and leucocytes yield very active 'preparations of fibrin ferment,' says Dr. Halliburton ; although he in no way proves it, it may be true. But if it should prove to be true that leucocytes, without any other organic addition, can be made to yield active preparations of fibrin ferment, this would not in the least alter or injure what I have said on coagulation.

I have said they do not contain it ; I have not said they might not, by various destructive processes, be made to yield it. On the contrary, I have always said that when they cause clotting in peptone-plasma they also cause the appearance of fibrin ferment. Whether any other form of destroying the cells may cause the formation of fibrin ferment I cannot personally say.

Dr. Halliburton seems to think that a leucocyte in a lymphatic gland alive and well is the same thing as a leucocyte treated with various salt solutions, dialysed, treated with alcohol, distilled water, and so on. It is, of course, easy to say, as Dr. Halliburton does, that leucocytes cause the clotting of the plasma because they contain a very active ferment. His experiments prove nothing of the sort. My own observations are quite opposed to this statement. We can only judge of the activity of a ferment solution by the rapidity with which it causes the clotting of a test fluid.

Now solutions of fibrin ferment which cause clotting in highly diluted strong magnesium sulphate plasma in two or three minutes cause, when added to peptone-plasma in the same proportions, no clotting whatever or only traces after hours. Yet leucocytes added to the same peptone-plasma cause clotting in a few minutes. If these experiments mean anything, they mean that the leucocytes act quite differently to strong fibrin ferment. And it was perfectly apparent from my first experiments, published in 1881, that the inter-action which occurs between the leucocytes and plasma was of a totally different nature to a fermentative process.

Now concerning Dr. Halliburton's observations I would say the following :—

They leave the chemical aspect of coagulation quite untouched ; his theory is therefore naturally simpler than mine. For to say that coagulation is due to a ferment is merely to envelop the whole question in the hopeless mist in which it has been for the last twenty years.

Dr. Halliburton says the white corpuscles liberate fibrin ferment when blood is shed, and he says his theory explains why the blood coagulates outside the vessels. This is, of course, the old theory of Alexander Schmidt ; a theory which has long been shown by its author and his pupils to be hopelessly untenable. It was shown to be impossible in my first papers in 1881.

Dr. Halliburton's repetitions of Schmidt's experiments and his deductions therefrom are as follows. By destroying lymph-cells in various ways he obtains solutions which he says contain the fibrin ferment. Therefore, he says, when the white corpuscles leave the blood-vessels they break up and yield this ferment. Finally, he says, leucocytes contain a very active ferment, a much more active ferment than serum.

Now if, for the sake of argument, we accept these observations as correct, it is clear that there is a very large gap between Dr. Halliburton's deductions and his observations, because we know with certainty that white blood-corpuscles when they leave the vessels contain no fibrin ferment. Leucocytes and white blood-cor-puscles are then, according to this, totally different, if Dr. Halli-burton's statements are correct and he means what he says. Under these circumstances, therefore, the ferment theory as propounded and maintained by Schmidt falls to the ground in Dr. Halliburton's hands. He himself does not appear to grasp the fact that Schmidt was quite aware that the leucocytes in the blood contained no fibrin ferment, and that he accordingly put forward the hypothesis that a specific form of protoplasmic change occurred which led to the development of ferment from the disintegrated corpuscles.

Now, unfortunately, the views of Schmidt and Hammarsten, edited by Dr. Halliburton, are supported by him by an experimental method which is open to fatal objections.

He uses, as a test for his supposed fibrin-ferment solutions, a fluid which he says clots rapidly with substances which are not fibrin ferment. It is, therefore, extremely difficult for him to know when he is dealing with the ferment and the non-ferment. He says that because his solutions cause clotting in his test plasma it cannot be concluded that they contain ferment, because other substances, e.g. muscle, cause rapid clotting. (In his first paper in the Royal Society he omits this.) It is, therefore, not at all clear how he knows he is dealing with fibrin ferment at all. His identification of the ferment would appear to depend on the temperature at which his solutions lose their coagulating activity, and even this is, according to his own statement, a somewhat precarious method of identification.

The fibrin ferment is, he asserts, 'cell-globulin,' a substance containing no phosphorus. 'It is a true ferment, an unstable substance, the product of a living cell, and that induces changes in the substances with which it comes in contact by a catalytic action—i.e. without itself apparently participating in the chemical changes its presence induces.' Dr. Halliburton, it may be seen, is very precise on the subject. But, as a matter of fact, all this rests purely on his *ipse dixit.*

Long before he took up the subject of coagulation it had been abundantly shown that all kinds of different substances when added to plasma cause coagulation. These substances were not fibrin ferment. And therefore it became of the utmost necessity to adopt the greatest precautions in testing for this substance. The two fluids which afford reasonable certainty that we are dealing with fibrin ferment are (1) the test fluid of Alexander Schmidt— i.e. highly diluted strong magnesium sulphate plasma, and (2) the fibrinogen of Hammarsten—i.e. the fibrinogen prepared from the strong $MgSO_4$ plasma of the horse by repeated precipitation and re-solution. They afford a reasonable certainty of the existence of ferment, because, so far as is known, they have little or no tendency to clot with 'other substances.' Dr. Halliburton appears very seldom to use these test fluids. As I have pointed out before, the serous fluids which he employs are very unreliable ; they vary in different cases, although as a general rule they resemble Hammarsten's fibrinogen in being easily affected by ferment and not by 'other substances.'

It must be remembered that the original and continued belief in the fibrin ferment depends almost entirely on our being able to separate it from proteids. Its recognition by Schmidt depended on this. The whole of Hammarsten's work depended on this. If it cannot be separated from proteid, then the proof that it produces 'changes without itself participating in the chemical changes becomes an exceedingly difficult one, and has never yet been presented to science.

Dr. Halliburton says 'that his ferment does not participate,' but I cannot see that he has afforded any material proof whatever. Indeed, he is not clear that his proteid solutions, which he calls 'ferments,' do not influence the amount of fibrin formed, and he affords no direct proof whatever that they are not materially altered. He does not give convincing evidence that the various substances he speaks of as 'fibrin ferment' are one and the same body. It is, of course, extremely hazardous to assert that one proteid is identical with another. But first there is one point Dr. Halliburton seems to have omitted to profit by. His cell-globulin from lymph-glands is 'quite free from phosphorus.' This is the first proteid, if this is true, ever found free from phosphorus (Bunge, Ritthausen). It is a pity he has not applied this method of identification to his other cell-globulins.

Then Dr. Halliburton makes a very remarkable statement about the fibrin ferment, which is entirely opposed to all previous experience. He says that even $\frac{1}{2}$ or 1 per cent. solutions of sodium chloride hinder very considerably the action of the ferment. I hope Dr. Halliburton has made direct observations on this point, since it is another of the sweeping and unsupported statements he has made concerning my work. The action of sodium chloride has been very fully worked out by the Dorpat school. The presence of salt solution up to 3 per cent. causes not a hindrance but a decidedly increased rapidity in fermentative coagulation. Against this well-known fact he brings no experimental evidence. The sole original point which Dr. Halliburton has made is that the fibrin ferment is a proteid. All the observations he gives on clotting have been made before. We have known for years that leucocytes from lymph-glands cause clotting, that extracts containing proteid from leucocytes cause clotting, that a proteid from serum causes clotting, and it is still for Dr. Halliburton to prove that these substances are the fibrin ferment, or that they act as an enzyme at all, and even then we shall not have got much further in the knowledge of the intimate nature of coagulation.

How extremely little value can be placed on the effect of heating as a test for an enzyme can be seen from an earlier observation I have long ago recorded. Tissue-fibrinogen causes intravascular clotting ; after the solutions have been boiled they lose this power. Yet they are just as active on extravascular plasma as on peptone-plasma.

In conclusion, then, Dr. Halliburton's sole experimental point against me is that leucocytes cause clotting in peptone-plasma because they contain a ferment so very much more active than the ferment of Schmidt or than serum. He does not, however, adduce one single observation which proves that leucocytes contain or even yield any ferment whatever, and there is no attempt to prove that their alleged ferment is so very active.

The observer he quotes in support of his views flatly contradicts him. Surely an observer in Dr. Halliburton's position, who brings such sweeping criticism to bear on my work, is bound to show that his one observation 'which completely upsets one of the very foundations upon which my theory rests' is a true and correct one. But at present Dr. Halliburton has only made the assertion, he has not attempted to prove it.

PART II

PATHOLOGICAL PAPERS

NOTE ON PROTECTION IN ANTHRAX [1]

HITHERTO in the few cases in which protection against zymotic disease has been found possible, it has been effected by the communication to the animal of a modified form of the disease against which protection is sought.

I have succeeded in protecting rabbits from anthrax by an altogether different process, and although this is scarcely, at present, of practical utility, it may perhaps be found to be of some interest as regards the general nature of protection in this and other diseases depending on micro-organisms.

I use as a culture fluid for the anthrax bacillus a solution of a proteid body which is obtained from the testis and from the thymus gland. I have described this substance to the Society on a previous occasion,[2] so that I need not repeat the description of the process used in its preparation.

The proteid substance is dissolved in dilute alkali and the solution sterilised by repeated boiling. It is then inoculated with anthrax and maintained at 37° C. for two or three days.

The growth is generally not very abundant, and at the end of the period mentioned is removed from the culture fluid by filtration. A small quantity of the filtered culture fluid is injected into the circulation of a rabbit, and it is then found that the animal will not take anthrax.

A subcutaneous inoculation of extremely virulent anthrax blood made at the time of the injection of the protecting fluid, and two subsequent inoculations at intervals of five and ten

---

[1] [From the *Proc. Roy. Soc.* vol. xlii. 1887.]

[2] 'Intravascular Clotting,' *Proc. Roy. Soc.* 1886. [Collected Papers p. 135.]

days, remain entirely without effect. The animals used as a control invariably die. Four rabbits have been protected in this way.

If the anthrax grown in the fluid be inoculated it either kills or it has no effect. It does not protect in the slightest degree.

The injection of the culture fluid in which no anthrax has grown is without effect. The animals die as usual when inoculated. The injection of the fluid itself causes no ill symptoms whether anthrax has grown in it or not.

If other albuminous fluids—e.g. blood serum—be used as a culture medium and the filtered culture fluid be injected, it exerts no protection. It may be fairly concluded that the growth of the anthrax bacillus in the special culture fluids used in these experiments gives rise to a substance which, when injected into the organism, protects against an immediate and subsequent attacks of anthrax.

It would obviously be of very great advantage if some such method as this could be used for the zymotic diseases affecting man for which no protective inoculation in the ordinary sense appears possible.

I am indebted to the Medical Officer to the Local Government Board for permission to publish this short account of these experiments, the full description of which will appear in his report. I must also express my thanks to Dr. Klein, F.R.S., for kindly supplying me with many anthrax cultivations.

Note added

The following experiments give additional weight to the previously described results:—

In the one case the anthrax grew with very great rapidity in the culture fluid, and the clear filtrate contained but a very small quantity of proteid matter. Forty cubic centimètres of this fluid were injected into a rabbit, and the rabbit immediately inoculated in the ear with virulent anthrax blood; in two days there was very marked œdema at the seat of inoculation, which increased to an enormous extent during the next few days, and then gradually subsided. The rabbit is now perfectly well, twenty-four days after the inoculation.

In the second case the growth of anthrax had been very slight; 20 c.c. of the filtered fluid was injected, and the animal immediately inoculated in the leg with virulent anthrax blood. In three days there was marked œdema at the seat of inoculation. This spread up the leg to the back, so that there was enormous œdema occupying nearly the whole posterior part of the animal; this persisted for ten days, and then gradually subsided. The animal is quite well, twenty-eight days after inoculation.

These cases are of interest, since they are obviously instances of partial protection. The animals are still affected by anthrax, but it is only as a severe local affection, and does not kill them.

PRELIMINARY REPORT ON THE MODE OF ACTION OF PATHOGENIC ORGANISMS [1]

THERE can be scarcely any doubt that the ultimate action of pathogenic organisms in producing disease is a chemical one; and, although suggestions and theories have not been wanting as to the nature of this action, the subject is still involved in very great obscurity.

Dr. Klein has placed at my disposal, for the purposes of research, a bacillus which produces a rapidly fatal form of septicæmia in guinea-pigs and rabbits. Dr. Klein has himself studied and described the morphological and physiological characters of this organism, so that it will only be necessary for me to give an account of the observations I have been able to make as to the mode in which this organism produces its fatal effects. It must be added that, owing to the short time I have been engaged on this subject, these observations are still very incomplete.

It has recently been shown [2] that, under the influence of certain non-pathogenic organisms, poisonous substances of the nature of alkaloids are produced in organic media; and it has been assumed that such poisons may arise in the animal body as the result of the activity of pathogenic organisms, thus accounting for the toxic influences of the latter. But satisfactory proof that pathogenic organisms are capable of producing such poisons does not at present exist; and as this is a very important subject, attention was particularly directed to this point in the case of the septicæmic bacillus now under consideration.

For an inquiry of this nature there are objections to the use

[1] [From the ' Report of Medical Officer,' Local Government Board, 1887.]
[2] Brieger, ' Ueber Ptomaine.'

of the more complex media more generally employed for purposes of culture, such as peptone-broth, or peptone and sugar; and in the case of this particular organism it was found more suitable to make use of a vegetable albumen. Vegetable albumens are obtainable commercially from various sources, and the one used in these experiments has been obtained from the Brazil nut.

The raw product is extracted with dilute alkali, filtered, and the clear solution precipitated by neutralisation with acetic acid. The solid matter is collected by means of the centrifugal machine and washed. It is again dissolved in dilute alkali, and this solution forms the cultivating medium for the septic bacillus under consideration. With regard to this solution, it is to be observed that it may be injected in large quantity into the circulation of the rabbit without producing any obvious disturbance.

Before inoculating the bacillus into the solution of vegetable albumen, the solution is thoroughly sterilised by prolonged boiling on several successive days. Provided sufficient alkali be present, the only change the albumen solution undergoes on boiling is that it becomes very slightly opalescent. It will be necessary hereafter to ascertain the exact amount of proteid and of alkali which the solution should contain; here it may be said that it is of great importance to avoid any excess of alkali, as this exerts an unfavourable influence on the growth of the bacillus.

When the solution of vegetable albumen has been fully sterilised, it is inoculated from a pure culture of the bacillus on peptonised gelatine broth, and the flask containing the albumen solution is kept in the incubator at a temperature of 36°–37° C. After a short period, varying from two to four days, a voluminous precipitate occurs in the fluid. The time which elapses between inoculation of the solution and the appearance of the precipitate depends on conditions not yet understood, among them the degree of alkalinity of the fluid probably being the most important.

When the precipitate has appeared, a small quantity of the contents of the flask is drawn off by means of a sterilised pipette, and from this gelatine tubes are inoculated, so that a control as to the purity of the culture may be obtained. The general characters of cultivations on gelatine are given in Dr. Klein's paper.

Microscopically examined, the precipitate is found to contain large numbers of bacilli, but at the same time it is found on careful examination that these bacilli do not make up the bulk of the precipitate. That consists principally of a very finely granular substance, which chemical examination shows to be chiefly proteid matter. Indeed, when the precipitate has fully developed, the supernatant fluid is found to contain only a very small quantity of proteid.

The culture fluid is still markedly alkaline when the precipitate is fully developed. Hence the precipitate cannot be referred to the production of an acid body by the bacillus. The growth, then, of the bacillus has caused the precipitation of the bulk of the albuminous matter contained in the original solution. This matter can hardly be regarded as representing the *débris* of organisms ; for, in the first place, the rapidity with which the precipitate is formed renders this almost impossible ; and secondly, the bulk of the growth and of the *débris* of growth which is formed in other fluid media (for instance, peptone-broth) is very much smaller than the precipitate under consideration.

The precipitate caused by the operation of the bacillus can be readily collected and washed by means of the centrifuge. When so collected it is found to be much less readily soluble in dilute alkali than was a precipitate formed by acetic acid from the original solution of vegetable albumen.

The bacillary precipitate having been treated with dilute alkali is filtered through a fine linen cloth, and in this way a fluid is obtained, turbid, but free from any visible solid particles. This fluid possesses very marked toxic properties, for if injected into the circulation of a rabbit it will cause death in less than a minute.

To produce this result, however, a considerable quantity of the fluid must be injected, or, more exactly stated, the fluid injected must contain a considerable quantity of the precipitate in suspension. I estimate the quantity necessary to kill a rabbit as 0·5 grm., but it must be understood that this is only an approximation. A less quantity produces no very obvious immediate effects; but as the animal is under the influence of an anæsthetic when the injection takes place, milder symptoms may be masked. The animal is killed before the anæsthesia is over.

The fluid which was separated from the precipitate, that is to say, the original culture fluid after the removal of the precipitate, does not exert any immediate toxic effect when injected in large quantity into the blood.[1]

No doubt the precipitate caused by the growth of the bacilli contains, after being taken up by the dilute alkali, a great number of the bacilli in suspension. But then, in like manner, there are a large number of bacilli to be found in that solution of the proteid which had shortly before been inoculated with them, and in which the precipitate has not yet made its appearance ; yet the injection of this solution is not followed by any immediate toxic effect ; so that the rapid toxic effect cannot be referred to the bacilli themselves.

The actual cause of the rapid death, above referred to, appears to be respiratory paralysis, the heart continuing to beat for some time after the respiration has stopped. In one or two instances I have observed intravascular clotting, but this is not constant. In four cases, however, a marked acceleration of the clotting of the blood after it left the vessels was noticed.

These changes in the blood are of interest because it has been shown [2] that a very poisonous substance can be obtained from many animal tissues. The substance is a complex proteid body, and it presents some analogies in its mode of action with the precipitate produced by our septicæmic bacilli. As has been stated, the latter consists chiefly of proteid matter ; whether it is this which has the toxic power, or whether this power is due to some other substance contained in the precipitate, must be left an open question till further experiments can be made.

---

[1] 'Immediate,' because the animal is killed before the expiration of the anæsthesia.

[2] *Proc. Roy. Soc.* Feb. 4, 1886. [Collected Papers, p. 135.]

One conclusion may be safely drawn from these observations, viz., that the growth of this bacillus in vegetable albumen solution does not result in the production of any soluble poisons, since only the precipitate has any toxic action—the fluid is inert. If it can be definitely shown that growth of bacilli in solutions of innocuous proteids can rapidly convert these into noxious bodies, a great advance in our knowledge of the action of pathogenic organisms may be possible, and perhaps these observations may be taken as an indication in this direction. They are certainly not favourable to the ptomaine theory.

In addition to the observations already described, one other point of interest was brought out. It was found that the organisms could be cultivated in the following fluid:—

Kreatin, 1 grm.

Glucose, 0·5 grm.

Potassium phosphate, 0·5 grm.

Lime and magnesia, traces.

½ per cent. solution of common salt, 100 c.c.

The growth produced was always of very limited extent, but the toxic influence of the organisms was preserved through many successive generations. But although the organism still retains its activity when inoculated, it is very much less virulent than when grown in peptone-broth or other albuminous fluid. Inoculations from a growth of peptone-broth cause death in guinea-pigs in from two to four days; whereas the average time elapsing between inoculation and death in the case of the growth in the kreatin solution was seven days.

RESEARCHES ON CHEMICAL PROTECTION [1]

IT is well known that in many zymotic diseases the blood is profoundly affected, and that in these diseases we meet with symptoms of general intoxication, associated with localised pathological processes, and with the remarkable phenomenon of protection against a subsequent attack of the disease. In the course of my investigations on the coagulation of the blood I have met with results so obviously related to the processes alluded to above as to inspire me with the hope of approaching more nearly the 'Chemismus' of zymotic disease.

I must express my thanks to my revered teacher, Professor C. Ludwig, for having allowed me to carry out many experiments, which are recorded in this paper, in his laboratory.

I. *The Action of Tissue-fibrinogen* [2] *on the Blood*

Every watery extract of any fresh tissue may be considered as a solution of tissue-fibrinogen. To purify the solution it is precipitated with acid. The precipitate is washed, and dissolved in very dilute alkali. The injection of this purified slightly alkaline solution into the vein of a rabbit produces total thrombosis of the whole vascular system. In the dog, on the other hand, thrombosis is confined to special regions, viz., the portal venous system, as I have already described in these 'Archives,' 1888, page 174.[3] The dogs recover, in many cases, from

---

[1] [Translated by the author from *Du Bois' Archiv*, 1888, p. 527.]

[2] Tissue-fibrinogen, *vide* 'Intravascular Clotting,' *Proc. Roy. Soc.* 1886; 'Ueber intravasculäre Gerinnung,' *Du Bois' Archiv*, 1886 ; 'Beiträge zur Frage der Gerinnung,' *ibid.* 1888 ; 'On Hæm. Infarct. of Liver,' *Brit. Med. Journ.* Nov. 1887. [Collected Papers, pp. 135, 253, 346.]

[3] [Collected Papers, p. 253.]

this lesion. But if, some hours later, a second injection of this solution be made, it has practically no effect, even if it be in much larger quantity. The first injection produces not only the constantly occurring local thrombosis, but the blood is so altered that it is rendered indifferent towards a second injection. This condition may last, according to the amount of the first injection and other circumstances, for several days. Thrombosis and diminished coagulability of the blood are quantitatively closely associated. If small quantities are injected, only insignificant thrombosis results, and the shed blood clots slowly. If large quantities are used, there is extensive coagulation, and the shed blood clots with the greatest difficulty. To illustrate the above I describe the two following experiments :—

1. Forty c.c. of a solution of very freshly prepared tissue-fibrinogen were injected into the jugular vein of a dog (weight 5½ kilos.). After an interval of from one to two minutes there was a very slight convulsion and the respiration stopped completely; after three minutes the respiration returned and continued for five minutes, then ceased permanently; no recovery of consciousness. Sec. cadav. was made at once. The portal vein and all its branches leading to the liver were completely thrombosed. Also most of the mesenteric veins, the splenic vein in particular, very strongly thrombosed. The lymphatic vessels leading from the liver were full of blood. Blood was effused beneath the peritoneal covering of the gall-bladder. The liver, which was otherwise pale, showed extensive and distinct hæmorrhagic infarctions. In the right heart was a fibrous shrunken coagulum, the animal being in full digestion.[1] Elsewhere there were no coagula.

The blood from the heart was collected, and not being spontaneously coagulable it was centrifugalised ; the plasma of this blood coagulated on further addition of tissue-fibrinogen, and also on the addition of leucocytes from lymph-glands. Left to itself, coagulation began in twenty-four hours.

2. Dog. Weight, 7 kilos. during digestion. At 11.40 A.M.

<hr>

[1] *Vide* these ' Archives,' 1888. [Collected Papers, p. 253.]

50 c.c. of the same solution were injected. Temperature of the dog before the injection 38° C. Immediately after the injection convulsions, cessation of respiration ; pulse in the cruralis not to be felt.

The animal recovered quickly and was let loose. After recovery from the (chloroform) narcosis, it was observed to have marked weakness of the hind legs for a considerable period. At 3.30 P.M. the animal was again narcotised (the temperature of the dog was 40·1° C.), and had 50 c.c. of a solution of the same tissue-fibrinogen, but of double strength, injected into the other jugular vein. A scarcely perceptible struggle occurred ; the breathing went on quietly. After five minutes the carotid is opened, the blood streams out under high pressure, and without difficulty 280 c.c. are collected. (On the other hand, as I have previously recorded, the result of a first injection is to cause so great a fall in blood-pressure that it is difficult to get a single drop of blood from the carotid.)

The animal was killed by chloroform. In the vena porta is a shrunken thrombus 2 to 3 cm. long ; the margins are already distinctly decolourised. The clot is continued upwards into several branches of the portal vein in the liver. It is prolonged downwards into a very thin white centrally-placed fibrin thread not thicker than ordinary sewing-cotton. Such threads are also to be found in the mesenteric veins also distinctly decolourised. The splenic vein contains a large shrunken and distinctly decolourised thrombus.

After very careful examination I found two mesenteric veins with fresh coagula not decolourised. The right heart contains a very small quantity of fibrous completely decolourised coagulum. The pulmonary artery, left heart, venæ cavæ, subclavian veins, iliac veins, contain completely fluid blood.

The liver shows unimportant hæmorrhagic spots ; the kidneys are microscopically unaltered ; the urine is coloured with methæmoglobin, due, I am inclined to think, to the rapid decolourisation of the clots. The other internal organs present no abnormality.

The blood was centrifugalised ; the plasma was slightly fatty,.

so that in large quantity it appeared whitish, but in small test-tubes it was quite clear.

This plasma has the following properties. It did not clot on addition of tissue-fibrinogen during six hours—time of observation. With leucocytes from lymph-glands no coagulation in six hours. With a very active solution of fibrin ferment no coagulation in five hours. (To 10 c.c. plasma 5 c.c. ferment solution added. Two c.c. of the same ferment solution produced coagulation in 25 c.c. dilute $MgSO_4$ plasma [ferment test] in fifteen minutes.)

With lecithin a rapid trace of coagulation occurred, but did not increase. That the plasma contains none of the injected tissue-fibrinogen is evident from the following : The solution of tissue-fibrinogen, even after it has been diluted twenty-five times, gives a permanent and distinct precipitate after addition of acetic acid in large quantity ($\frac{1}{2}$ c.c. acetic acid 35 per cent. to 2 c.c. fibrinogen solution). The plasma treated with acetic acid in the same proportion remains perfectly clear. The plasma remained entirely fluid for three days.

Nevertheless, the plasma contains much fibrinogen. With dilute sulphuric acid (0·4 per cent.) added till a marked acid reaction is obtained, it gives a voluminous proteid precipitate. In another portion of the plasma the fibrinogen is precipitated in the usual way with NaCl. The precipitate is washed with saturated NaCl solution till disappearance of proteid reaction, the salt removed, and the precipitate dried and weighed. I found 0·93 per cent. fibrinogen.

If it be remembered that the great mass of the blood has remained fluid, and that only in the portal and splenic veins extensive thrombi have formed, the large amount of fibrinogen still left in the blood is in no way surprising.

The point of interest is that, although the solution is *added to the whole circulating blood, it only produces pathological changes in certain definite regions, and that the first injection so changes the fibrinogen of the blood that it remains indifferent to a second injection, both when in the vessels of the living animal and when shed.*

II. *Use of Boiled Tissue-fibrinogen as Culture Fluid. Experiments with Anthrax*

In order to study the changes which solutions of tissue-fibrinogen undergo owing to the action of pathogenic organisms, it appeared to be of importance to use a bacillus which acts quickly and certainly. I have, therefore, confined my observations chiefly to the bacillus anthracis.[1]

The solution is always sterilised by boiling before being inoculated; and as the result of boiling, the fluid undergoes changes in its chemical and physiological characters which I will first describe.

The boiled solution has lost its power of producing coagulation when injected into the circulating blood, but it retains the power of producing coagulation in extravascular plasma,[2] *e.g.* peptone-plasma. As the result of boiling, a greater or less quantity of the tissue-fibrinogen is coagulated. The coagulum cannot, however, be regarded as ordinary coagulated albumen ; for the coagulated (by boiling) fibrinogen possesses in a high degree the power of inducing coagulation (fibrin) in peptone-plasma. With regard to the quantity and properties of the coagulum, different solutions are in no way identical.

We can conveniently describe three varieties.

Either the greater part of the fluid coagulates, or the fluid becomes merely opalescent, or it is not appreciably changed.

The different behaviour depends certainly in part, but not exclusively, on the amount of alkali present in the solution.

By means of a certain degree of alkalinity, solutions can always be obtained which are not changed by boiling. But the amount of alkali requisite to produce this result cannot at present be generally stated, because tissue-fibrinogen is a very changeable body, and because the extracts which are obtained from different specimens of thymus are not precisely identical. For convenience I will style such solutions, which undergo no change on boiling, ' strong alkaline solutions.' It must be

---

[1] *Vide* ' Report of Medical Officer,' Local Government Board, **1887**, and forthcoming Report. [Collected Papers, pp. 324, 340.]

[2] *Ibid.*

understood that this term is used relatively—*i.e.* sometimes such solutions are really very alkaline, sometimes only weakly alkaline.

The different behaviour of different boiled solutions of fibrinogen is a marked feature in their behaviour as culture fluids. In the 'strong alkaline solutions' the bacilli grow with very great rapidity and in great abundance, and they kill, after many days' culture, with great rapidity when inoculated. But the fluid itself is not poisonous. It can be easily separated by ordinary filter-paper from the bacilli, which form a coagulum-like mass of long fibres. The filtrate can be injected into the circulation of a rabbit without producing any evil effect whatever. The animal acquires no immunity as the result of the injection. It dies in the usual period if it be inoculated subcutaneously with anthrax blood.

'Weak alkaline solutions' behave quite differently. In these the anthrax bacilli sometimes will not grow at all, or they grow to a certain extent and lose their virulence, or they rapidly exhaust the proteid of the fluid and are still very poisonous.

With such 'weak alkaline' culture fluids I have been repeatedly able to render rabbits immune as regards anthrax; but the bacilli must not be injected into the blood together with the fluid.

They can be separated from the fluid by filtration, as has just been described. This procedure, however, with scanty growth, is uncertain, and it is better to kill the bacilli by boiling. But now it is found that the fluid—which, as described, has already been boiled previous to inoculation—has again become sensitive to boiling, and displays a greater or less tendency to form an additional coagulum. This point will be referred to subsequently. If the solution, now freed from bacilli, be injected into the vein of the animal, the latter as a rule acquires immunity towards anthrax—*i.e.* it can, immediately or later, be subcutaneously inoculated with the most virulent anthrax blood without any ill effects.

I have protected a large number of animals in this way, so that there is no doubt about the observation being correct.

The protective action lasts very long.   In one case it was in full power after an interval of fifteen months.

The best way to proceed to obtain protection is as follows : The watery extract of thymus, or the very weak alkaline solution of the acetic acid precipitate, is boiled, diluted, and filtered through linen.   The fluid is inoculated with anthrax, and the culture incubated for two or three days.   Now the fluid is boiled to kill the bacilli.   Should the fluid display a tendency to form extensive coagula on boiling, alkali must be added.   After boiling it is filtered through linen, and it is now ready to inject.

With such a fluid I have been able in one series to protect eight out of nine rabbits used.   In the case of the ninth, the proteid coagulated so firmly on boiling that it would not filter ; consequently the fluid transfused was almost proteid free. Cultivations in which the tissue-fibrinogen so completely coagulates on boiling that clear filtrates are obtained giving no proteid precipitate with acetic acid are useless.   The injection of such solutions has no influence on the course of inoculated anthrax.

I record here two observations which have been made for special purposes.   They will serve as illustrations for describing in detail the procedure.

1. November 10.—The watery, very feebly alkaline extract of a thymus yields on boiling an extensive coagulation.   The filtrate is opalescent, and contains abundant proteid easily precipitated with acetic acid.   The fluid is sterilised and inoculated with anthrax.   After three days in incubator, the anthrax settles as a coagulum-like mass at the bottom of the vessel.

The fluid above is removed, filtered, and boiled.   Two rabbits are inoculated with this fluid.

Rabbit I., of 4 lbs. weight, has 30 c.c. of the solution injected into the jugular vein.   Thereupon it is inoculated with blood from the heart of a guinea-pig dead of anthrax.

Rabbit II., of 2½ lbs. weight, has 25 c.c. of the solution, in several portions, injected subcutaneously.   Then it is inoculated with the same blood as Rabbit I.

A guinea-pig is simultaneously inoculated with the same blood.

November 12.—Guinea-pig found dead.

November 13.—Rabbit II. found dead.

November 16.—Rabbit I. is quite well, and is again inoculated with anthrax. Six months later still alive.

From this it must be concluded that the *subcutaneous injection of the fibrinogen does not protect.*

2. April 28.—The watery extract of a thymus gives a distinct coagulum on boiling. After addition of water it is filtered through linen. The strongly opalescent filtrate, after repeated boiling, is inoculated with anthrax. After two days in incubator, the culture—in which the bacilli have grown abundantly—is, without boiling, filtered through paper. The fluid is at first not clear, the pores of the filter soon become stopped, and the filtrate becomes clear; consequently the filter is frequently changed in order to prevent the filtrate becoming too poor in proteid. The collected filtrates are strongly opalescent and *contain anthrax bacilli.*

Without previous boiling, 40 c.c. are injected into the jugular vein of a rabbit. Simultaneously it is inoculated in the ear with anthrax blood.

May 1.—The animal is dead. No œdema of the ear, but œdema of skin of abdomen. Blood full of bacilli.

Thirty c.c. of the solution are boiled for several minutes and filtered through linen. The filtrate is injected into the blood of a second rabbit. Immediately afterwards the animal is inoculated with anthrax blood. With the exception of a very slight redness at the point of inoculation, there is no result. A second inoculation is just as ineffectual.

It would appear that the fluid only protects from subcutaneously inoculated anthrax.

III. *Changes of Tissue-fibrinogen by Boiling. Protection without Anthrax*

The observations recorded above appear to me to give some insight into the nature of protection. At the present time the tendency is to believe that pathogenic organisms are injurious on account of the poisons which they produce, and it can hardly be doubted that they are able to form alkaloid-like substances in certain culture fluids. But it may well be doubted whether they are able to produce such substances in the living tissues of the organism. No artificial fluid can be regarded as precisely identical with the living tissues. But it must be borne in mind that from almost all tissues substances can be obtained which are closely allied to the fibrinogen of the blood, and coagulate with the latter when it permeates a diseased vascular wall or escapes as the result of a breach of continuity. Since I had found that solutions of tissue-fibrinogen still maintain, in part, their physiological activity on boiling, it appeared to me that such a fluid was more suited for obtaining information than the usual peptone, &c., solution.[1]

Now one very marked result of my observations is that a very abundant growth of anthrax in a fibrinogen solution confers no poisonous properties on the fluid. It can be injected into the blood in large quantities without any ill result.

With appropriate precautions this injection leads to the production of immunity, and it might be imagined that the bacilli separate a ' protecting substance.' But the facts do not warrant such an assumption.

First, the protective influence is not at all proportional to the growth of the bacilli. On the contrary, the cultures in which the bacilli have grown with great luxuriance are generally entirely inert.

Secondly, I have found that, in order to protect, the *solution injected must contain a certain quantity of fibrinogen, probably always in an altered state.*

It would therefore appear that the action of the bacilli depends on their producing a special modification in the proteid of the solution.

That the growth of the bacilli changes the character of the fibrinogen may be gathered from the foregoing, and this point may be illustrated by a special experiment.

A chopped-up thymus is treated with distilled water. After twenty-four hours the mixture is strained, and a few drops

---

[1] *Vide* ' Report of Medical Officer,' 1887. [Collected Papers, p. 324.]

of NaHO added. The reaction is just alkaline to neutral test-paper. It was boiled, and directly after boiling treated with an equal quantity of distilled water, and filtered through linen. The filtrate is opalescent, so as to appear milky. After addition of one or two drops more alkali, the fluid is put in a sterilised flask and repeatedly boiled. It is inoculated with anthrax and incubated two days. After this time a fairly abundant growth can be observed under the microscope as long threads.

The reaction is still just alkaline, but on boiling a portion the whole of the proteid coagulates to a solid mass. Addition of small quantities of alkali does not prevent this. Not till the reaction is made strongly alkaline does a part of the proteid remain in a swollen-up state of semi-solution.

The anthrax growth changes a solution of tissue-fibrinogen in such a way that it again becomes sensitive to boiling. A very moderate growth of anthrax suffices to effect this change. Bearing in mind the extremely unstable character of fibrinogen, it occurred to me that it was by no means out of the question to obtain the desired modification without the intervention of anthrax.

I have already alluded to the fact that solutions of tissue-fibrinogen from different sources are seldom identical. Another factor of great importance is the procedure which is adopted in the filtration of the boiled and partly coagulated culture fluid.

This can be illustrated by the following :—

A well-chopped thymus is allowed to stand for twenty-four hours with distilled water. The fluid is centrifugalised and at once boiled without any addition. The fibrinogen passes into the coagulated form partly as flocculent shrunken masses, partly as a swollen mass which makes the fluid opalescent. By the addition of small quantities of alkali the firmly coagulated part is diminished, the part remaining in semi-solution increased. If such a solution be filtered through ordinary filter-paper, the first filtrate is turbid and contains fibrinogen ; the latter is precipitated as soon as the fluid acquires a distinctly acid reaction. But the pores of the filter rapidly become stopped, and the later portions of the filtrate are clear as water. They contain small quantities of proteid differing from the first fibrinogen precipitate.

It is therefore evident that, according to the fineness of the filter, the degree of semi-solution, the quantity and character of the firm coagulum, the filtrate of a boiled fibrinogen solution may be of very varying character, so that it is very difficult to obtain a standard solution.

For protection purposes, the problem is to establish a certain change in the fibrinogen by boiling. For this purpose the quantity of alkali present must be extremely small, as otherwise the change is certain not to occur. On the other hand, the fibrinogen must not be converted wholly into the firmly coagulated variety, as it is then not in a state in which it can be used for experiment.

In my first experiments I was unaware of many of the conditions which might influence the culture fluid, and finding culture fluids in which anthrax had grown possessed the power of giving immunity, whilst those in which no anthrax had grown at most prolonged the period elapsing between inoculation and death, I was inclined to regard the anthrax bacilli as necessary factors in the matter.

But since it has become evident that the protection depends on a particular modification of fibrinogen, and since the most varied solutions can be obtained according to the physical processes of preparation, the necessary intervention of anthrax is not certainly established.

On the contrary, I have succeeded, after many failures, in protecting two rabbits by the injection of simple fibrinogen fluid—*i.e.* in which no anthrax had grown.

In these two cases the fibrinogen was injected in a rather strongly coagulated state—*i.e.* the boiled fluid, which coagulated, was pressed with force through linen, so that a considerable quantity of actually coagulated fibrinogen in a very finely suspended form was injected.

I hope shortly to be able to give further information on this subject. A connected statement like the above appeared to me desirable, as helping to render a very obscure and difficult subject more intelligible.

FURTHER REPORT ON THE MODE OF ACTION OF PATHOGENIC ORGANISMS [1]

DURING the past year I have continued my investigations of the mode of action on the animal body of pathogenic organisms. The method adopted in these researches has sought, in the first place, to establish, with regard to particular pathogenic organisms, some definite physiological data. In this way it is hoped that indications may be got of related chemical problems of more immediate practical interest, and the scope of the chemical side of the inquiry become correspondingly defined ; for it is obvious that the chemical problems involved in the mode of action of pathogenic organisms are so wide of range and of so obscure a nature that it is absolutely essential for an individual investigator to define, at any rate at the beginning of his inquiry, the limits within which he is going to work.

For experimental purposes, for instance, it is of the highest importance to work with organisms which produce their effects with certainty and rapidity. Hence, the bacillus anthracis has this year been the chief subject of inquiry.

I have used as a culture fluid for this bacillus a solution of a proteid body which is most conveniently obtained from the testis of bulls and the thymus of calves or young oxen. This proteid body I have elsewhere [2] described. The most characteristic property it possesses is its power of causing widespread intravascular clotting when injected into the blood-current of a living animal. I cannot assert that the substance prepared from testis and that from thymus are absolutely identical ; but they so

---

[1] [From the ' Report of the Medical Officer to the Local Government Board,' 1888.]

[2] *Proc. Roy. Soc.* 1886.

nearly resemble one another in their chemical and physiological reactions that for present purposes I regard them as identical, and shall call them, therefore, ' fibrinogen of thymus and testis.'

Substances very closely allied to this fibrinogen can be prepared from very various tissues, and in our dealing with pathogenic organisms we have always to bear in mind that these organisms exert their special influence on *tissues*. We cannot, of course, obtain an artificial solution which exactly resembles any tissue in its chemical characters; but since we find in almost every tissue substances closely allied to the fibrinogen used in these experiments, and since fibrinogen retains in part some of its original physiological properties after boiling, it occurred to me that it might be taken as much more nearly resembling the soil in which the organisms normally develop their peculiar action, than the ordinary peptone-broth solutions, which certainly do not present the same resemblance in their chemical constitution to the tissues and fluids of the animal body.

As above stated, if a solution of fibrinogen be injected into the circulation, it causes very widespread intravascular clotting. It also causes very rapid clotting in artificial extravascular plasma. But a solution *which has been boiled* does not, when injected, cause clotting, nor produce any ill effects. It still retains the power of causing very rapid clotting in extravascular plasma— that is to say, plasma from blood which has by appropriate means been artificially prevented from clotting ; after boiling, therefore, it still retains in part its physiological properties.

Preparation of the fibrinogen culture fluid.—The organ, testis or thymus, is finely minced and mixed with distilled water, to which (for antiseptic purposes) a little chloroform has been added. This mixture is allowed to stand twelve to fifteen hours. The watery extract is then filtered off through linen, and the filtrate put on the centrifugal machine till no further deposit appears. The extract is then acidified with acetic acid ; this gives a bulky flocculent precipitate, which is the fibrinogen. The precipitate is washed to free it from the acid, and is then dissolved in dilute alkali and boiled. The more rapidly the total processes of preparation have been carried out, the more readily does the substance dissolve in alkali and the less alkali is required to prevent its coagulation when boiled.   On the other hand, if it has stood as a precipitate under water for some time, it requires more alkali to dissolve it, and is much more prone to coagulate on boiling. It will be understood from this that practically there is very considerable variation in the culture fluid, some samples containing more alkali than others, some more proteid.   Moreover, it may be assumed as certain that the influence which boiling exerts on the fibrinogen solution is considerably modified by the amount of alkali present.

The readiness with which anthrax grows in different specimens of the culture fluid is very variable, and this is not difficult to understand in view of the above statements with regard to the behaviour of the fluid.   I am still engaged in investigating the exact method which should be followed so as to produce a fluid of uniform character, and best suited to the purpose.

The alkaline solution of the fibrinogen is put into a sterilised flask and boiled from ten to fifteen minutes on two successive days ; it is then inoculated with anthrax and kept in the incubator at 37° for two or three days, that is, for a period which suffices for the formation of the definite growth in the form of flocculi visible to the naked eye.   This period may be a little under or a little over the time mentioned.   Before giving details of the experiments I will shortly state the results which have been obtained.

1. *When a definite growth of anthrax* has taken place in the fibrinogen solution the latter is *filtered off from the growth,* and about 20 to 30 c.c.[1] of the filtrate are injected into the jugular vein of a rabbit.   This causes no ill symptoms in the rabbit, but *protects the animal* from an immediate and from subsequent inoculations with anthrax.

2. Solutions of fibrinogen, either from testis or from thymus, which have been treated in exactly the same way as the culture

---

[1] The injections into the jugular vein are made in the following manner : The animal is etherised, a cannula tied into the jugular vein, and connected with a burette containing the fluid, and this is slowly allowed to run into the vein.   Precautions against the entry of any air are, of course, taken.

fluids described above, *but which have not been inoculated with anthrax, do not protect rabbits* when injected in like quantity into the jugular vein.

It is, as a rule, not difficult to obtain a clear filtrate of the culture fluid in which the most careful microscopical examination fails to detect any bacilli; and for this and other reasons, which I postpone to a later period, it is certain that the protection which the injection affords is of a chemical nature, and not a protective inoculation in the ordinary sense of the term.

The following are examples of experiments described above, and they have been made to include one or two observations carried out during April 1887 :—

1. *Anthrax grown in fibrinogen from testis.*—After two days in incubator there was abundant growth of threads of bacilli. The filtrate from this growth was injected into the jugular vein of a rabbit, the amount of fluid injected being 30 c.c. The date of the injection was February 2. At the time of the injection the animal was inoculated subcutaneously in a distant part of the body with blood from the heart of a guinea-pig dead of anthrax; this blood was full of bacilli, and its virulence was proved by inoculating a control rabbit, which died of anthrax after the usual interval. No symptoms followed the injection or inoculation of the rabbit first referred to. On February 6 it was again inoculated with anthrax blood, but no effect was produced. On the 11th it was inoculated a third time with anthrax, but without any result. All the animals used in each instance for control experiments died.[1]

2. *Anthrax grown in fibrinogen from thymus.*—After one day in the incubator there was a distinct growth. Of the filtrate, 30 c.c. were injected into a rabbit on February 24. At the same time the animal was inoculated subcutaneously with anthrax blood. No effect was produced. Again inoculated with anthrax on March 6, but again without effect. The control animals died of anthrax in the usual time.

3. *Anthrax grown in fibrinogen from testis.*—After four days

[1] The control animals—guinea-pigs and rabbits—died of anthrax before the end of the second and third day respectively.

in incubator growth had taken place very slowly, but there was a distinct growth.   March 31, 30 c.c. of the filtrate were injected into a rabbit, and a subcutaneous inoculation of virulent anthrax blood was made simultaneously.   No effect was produced.   Two subsequent inoculations with anthrax blood, on March 5 and 14, remained without effect.   The control animals died of anthrax.

Although in these experiments the amount of fluid injected was 30 c.c., I have gathered from later observations that a less quantity is equally efficient.

I think it may be concluded from these experiments that the growth of the anthrax bacilli in these solutions gives rise to a chemical substance or chemical substances, which, when injected into the circulation, confer immunity against anthrax.

I have reason to believe that the anthrax bacillus induces a special change in the fibrinogen, and that it is this modified fibrinogen which effects the protection.   It is quite possible that a similar modification of the fibrinogen might be effected by other means than the growth of the anthrax bacillus.   I have, in fact, observed that the injection of a boiled solution of fibrinogen which has not been inoculated with anthrax may prolong the period which intervenes between the inoculation of the rabbit with anthrax and its death, and this to considerably beyond the usual period.   But this result is inconstant, and the animals have always died of anthrax within a week.

In comparing the effects of a solution in which anthrax has not grown with one in which anthrax has grown, care has been taken that the treatment of the two solutions and amount injected in either case should be as nearly as possible identical.

I have two further experiments to record which may perhaps be regarded as of very special interest.   They were not concluded by the end of March, but I regard them as of such importance that I give them in this report.

4. *Fibrinogen from thymus* was prepared in the manner of which I have given a general description above, and was used as a culture fluid for anthrax.   The duration of stay in the incubator was three days, but the growth was extremely scanty. On April 1, 30 c.c. of the filtered fluid were injected, and at the

same time the rabbit was inoculated with virulent anthrax blood. On April 3 there was marked œdema over the whole posterior part of the animal.   On the 20th the œdema had disappeared, and on the 28th the animal having become apparently quite well, it was re-inoculated with virulent anthrax blood. This produced no effect, and the animal is still alive—twelve days later.

5. In this experiment also *fibrinogen from thymus* (prepared as before by precipitation, re-solution, and boiling) was used as culture fluid. The growth of the bacillus was extremely rapid, and it caused a precipitation of proteid matter, so that the filtered fluid contained very little proteid.   Of this fluid 40 c.c. were injected, and the rabbit at the same time inoculated with anthrax blood in the ear.   In three days there was œdema, which increased greatly, and lasted about ten days.   On April 30, the animal being quite well, it was re-inoculated with anthrax blood, which had no effect; the animal was quite well ten days later, and was still living at end of May.

The fluids injected, then, in these experiments 4 and 5, availed to restrict the operation of the inoculated anthrax to one part of the body, to prevent death by anthrax, and to leave the animal protected against virulent anthrax subsequently inoculated.   I am inclined to suppose that the different effects produced by the injection in experiments 4 and 5 is due in the one case (4) to the very slight growth of anthrax, and in the other (5) to the very small quantity of proteid matter injected; but this is a question which is still occupying me, and for this report my experimental data are too small adequately to discuss the question.   The fact that the growth of anthrax in the culture fluid is sometimes extremely slight, whilst nevertheless the latter acquires the power of producing protection, has made me doubt whether the intervention of the anthrax bacillus is a necessary factor.   Whilst admitting the possibility that a fluid might be obtained which would, without having been the soil for the growth of anthrax, protect, I can at present only say that I have been unable to obtain this fluid.

These appear to be very remarkable results as bearing on the question of cultivating protective fluids outside the animal body.

ON HÆMORRHAGIC INFARCTION OF THE LIVER [1]

HITHERTO it has not been possible to obtain experimentally a hæmorrhagic infarction of the liver. The branches of the portal vein are strictly analogous to an end artery, but owing to the fact that the hepatic artery breaks up into capillaries which anastomose with those of the portal vein, the occlusion of the branches of the latter by emboli is not followed by any hæmorrhagic infarct.

Thus Cohnheim and Litten found that large quantities of finely divided inert bodies could be injected into a mesenteric vein without producing any hæmorrhagic infarction of the liver, although the obstruction in the portal vein was such that ascites was set up.

In human pathology a simple hæmorrhagic infarction is, I believe, a rare or almost unknown occurrence. In the present note I shall describe an experimental method of inducing the most widespread infarction of the liver, and I hope to show that some of the commonest pathological changes in the liver possibly originate by this or an allied process.

The experimental process to which I allude consists in injecting a small quantity of a solution of a substance into the jugular vein of the dog. This substance is a complex proteid body which can be readily obtained from various tissues, particularly the thymus gland. It can, however, be got from many other sources, as liver, brain, testis, &c. I have described [2] the preparation and properties of this substance elsewhere, and need not repeat them here.

---

[1] [From the *Transactions of the Pathological Society of London*, 1888.]
[2] *Proc. Roy. Soc.* 1886. [Collected Papers, p. 135.]

If a solution of this substance be injected into the jugular vein of a rabbit the animal dies instantly, owing to complete intravascular clotting of the blood. If, however, it be injected into the jugular vein of the dog it only causes clotting in the portal vein, its tributaries and its branches—that is to say, if some of the substance be injected into a dog's jugular a clot is invariably formed in the portal vein. The amount of clot formed varies with the amount of substance injected. If the latter be large the whole portal system, from the finest divisions of the mesenteric veins to the terminal branches of the portal vein in the liver, is completely thrombosed. In these cases, where a large amount is injected, the animal generally dies at once, and frequently a small amount of clot is found in the right heart.

But by injecting smaller quantities it is readily possible to induce a very extensive thrombosis of the portal vein without killing the animal and without producing any appreciable clot anywhere else. If the animal·be killed a few days after the injection a clot is always found in the portal vein, and the liver shows more or less extensive infarctions.

This is the method. I will now describe, first, the early changes in the liver, and subsequently the later. The early changes are hæmorrhagic in character, and since they occur as the result of a process of thrombosis they may be classed as hæmorrhagic infarction, although, as I shall discuss later, it is questionable whether the process is identical with the ordinary infarction that one gets in the liver or spleen. The hæmorrhages partly occur in well-defined more or less pyramidal areas which have a deep red colour, and contrast strongly with the pale colour of the rest of the liver.

A microscopical section shows the capillaries of this area to be enormously distended with blood, whilst large actual hæmorrhages, both in the substance of the lobules and in the portal canals, are seen. But in addition to these patches, or in place of them, a diffuse hæmorrhage all over the surface of the liver may be met with, and the blood may be collected into large blebs beneath the capsule of the liver. A very peculiar fact must here be noticed with regard to the gall-bladder. It is especially common, after injection of even very small quantities of the proteid, to find most extensive extravasation of blood beneath the peritoneal covering of the gall-bladder. I noticed in one case in which this was very marked that the gall-bladder itself was filled with a colourless mucus of the most extreme tenacity; the mucus was just passing into the common duct, which it must inevitably have obstructed. It is possible that in this way slight changes in the blood may give rise to jaundice. I have noticed this hæmorrhagic affection of the gall-bladder as the result of the injection into one dog of blood-plasma from another dog. It may be noted that ascites is never produced as the result of the plugging of the portal vein, no matter how complete the latter may be.

The later changes in the liver as the result of this injection are as follows: In the first place the clot in the portal vein has always completely disappeared after an interval of fourteen days and probably much less. After this interval there is not the slightest trace of a clot to be found either in the main trunk of the portal or its branches. The liver may be apparently but little altered, except for some slight thickening of the capsule here and there in patches. Careful microscopical examination of the livers always reveals, however, scattered patches of early cirrhosis.

In other cases the whole liver shows the most profound structural changes—in an earlier stage large patches of what I regard as coagulation necrosis of the liver tissue; in a later, the most extreme fatty metamorphosis, together with less marked cirrhosis.

The very striking difference between my results and those of Cohnheim and Litten is only, I think, capable of the following explanation. I may first mention as a preliminary that no thrombus is discoverable in the hepatic artery as the result of the injection. We speak at present of two kinds of emboli or thrombi, simple and infecting. The thrombus produced in the portal vein by the method described is neither a simple inert thrombus, for the latter produces no infarction, nor is it infective, for it produces no abscesses. Its specific character lies in

PLATE 2.

To face page 348.

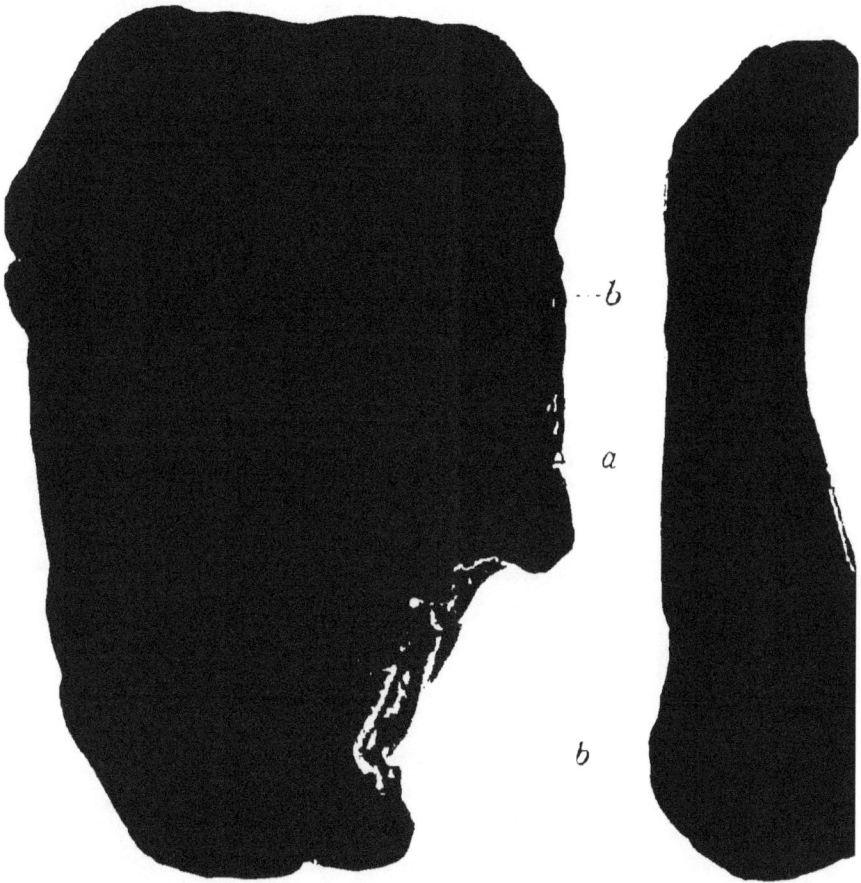

Fig. I.

Fig II.

the fact that it gives rise to most extensive hæmorrhages. The injection of the proteid substance used in these experiments not only causes clotting, but also induces very marked chemical changes in the blood. In fact, there is set up a temporary and mild sort of hæmorrhagic diathesis. The blood which is drawn off after the injection does not clot, or only very slowly and imperfectly. Where there is a local injury extravasations of blood are liable to occur. Thus there is always considerable hæmorrhagic infiltration of the tissues in the neighbourhood of the wound, which latter is of a most trivial character. I think it is the coincidence of these two factors, the hæmorrhagic condition of the blood and the thrombosis, which produces the infarction of the liver. As described above, the later changes in the liver are principally those which we are accustomed to speak of as chronic inflammatory changes, and perhaps these observations may throw some light on the origin of these very obscure processes.

This paper is only intended as a preliminary communication. A complete discussion of the observations recorded here would, of course, require much more of the time of the Society than I should feel justified in taking up. Specimens of the clot in the portal vein, of the infarction of the liver, and of the cirrhosis and fatty changes are shown.

November 1, 1887.

ON AUTO-INFECTION IN CARDIAC DISEASE

IN 1886 I described to the Royal Society [2] a substance, one of the most noticeable features of which was that it caused intravascular clotting when injected into the circulation of an animal. In subsequent publications I have further described the action of this substance, or rather group of allied substances, and speak of them as 'fibrinogens.'

In particular, I pointed out in my papers in Du Bois-Reymond's ' Archiv,' 1886, and in Ludwig's 'Festschrift,' 1887, that the lymph and chyle contained this substance. More exactly, I had found that the fluid of lymphatic glands, freed from all form-elements, possessed precisely the same action as the fibrinogens, and that the fibrinogen was the active substance in this fluid. The lymph contained in serous cavities does not contain this body, hence it is probably formed in the lymphatic glands. Dr. Krüger,[3] assistant to Professor Alexander Schmidt, of Dorpat, has disputed the correctness of these observations. But I am absolutely certain, from a repetition of my experiments, an account of which I have published elsewhere,[4] that Dr. Krüger is in error, and that my original observations were correct.

In the present paper I endeavour to show the light which further experiments have thrown on this question, and to point out the probably great importance which fibrinogen intoxication plays in a large and important class of disease, particularly cardiac disease.

---

[1] [From the *Proc. Roy. Soc.* vol. xlv. p. 309, 1889.]

[2] 'On Intravascular Clotting,' and ' Croonian Lecture Abstract,' April 8, 1886. [Collected Papers, pp. 135, 141.]

[3] Krüger, *Zeitschrift für Biologie*, 2 Heft, 1887.

[4] 'The Nature of Coagulation' (pamphlet, London, 1888). [Collected Papers, p. 201.]

For the purpose of my experiments I have used mainly the thymus gland, as the fluid and the fibrinogen of the thymus are quite similar to those of lymphatic glands, and are more easily obtained.

Exp. 1.—The $\frac{1}{2}$ per cent. NaCl fluid of the thymus, perfectly fresh, the cells completely removed by the centrifuge. The fluid rendered faintly alkaline with Na_2CO_3.

Dog I.—Weight, 19 lbs. Injected rapidly into the jugular vein 8 c.c. of the fluid. Dog killed. The portal vein was thrombosed, the clot commencing in the middle of the portal trunk, and extending into all the branches of the portal in the liver.

Dog II.—Weight, 16 lbs. 7·5 c.c. of the fluid injected, but ten times diluted with alkaline salt solution. The injection was slow, taking from three to four minutes. The dog was killed. There was absolutely no trace of clotting in any vessel.

As regards diet the animals were in similar conditions.

Exp. 2.—Used the watery extract of thymus, precipitated with acetic acid, and the solution of this precipitate in alkaline $\frac{1}{2}$ per cent. NaCl injected.

Dog I.—Weight of dog, 14 lbs. Injected rapidly 7 c.c. of solution. The animal ceased to breathe instantly and never breathed again. The heart continued to beat for several minutes. The right heart, the whole of the pulmonary artery and veins, and the left heart one solid clot.

Dog II.—Weight of dog, $13\frac{1}{2}$ lbs. 7 c.c. of the same solution injected, but diluted ten times with alkaline salt solution ; the injection slow, occupying three to four minutes. Dog killed. Absolutely no trace of clotting anywhere.

It is seen from the above experiments that a substance added rapidly to the circulating blood produces a pronounced effect ; added comparatively slowly and diluted, but in the same quantity proportionate to the weight of the animal, it produces no effect at all.

The obvious effect may be local—*i.e.* occur where the sudden admixture of fluids takes place, *i.e.* in the heart ; or it may be remote, and take place in the portal vein.

The phenomenon appears to resemble somewhat the so-called' ' mass influence' (*Massenwirkung*) of chemists.

A sudden admixture of a sufficient quantity of this substance with a given quantity of blood poisons the blood; the same conditions would be produced if, instead of the injection being sudden, the blood were circulating more slowly. In this case, also, a given quantity of the blood would in a given time receive a larger quantity of the fluid than if the blood were rapidly circulating. For the present I am speaking of the blood being affected by its showing an obvious change, that is, clotting; and I know, from previous experiments, that to produce this change a certain quantity of the fibrinogen must be added to the blood —*i.e.* the larger the dog, and consequently the more blood, the more of fibrinogen must be injected.

The present experiments show that to affect the blood a certain quantity of the substance must reach the blood within a given time, and this effect may obviously be obtained either by rapid injection or by the current of blood being slow in the neighbourhood of the vessel used for injection. I am therefore inclined to explain the fact that the lymph does not normally poison the blood because it runs into the blood slowly whilst the blood circulates rapidly. In a normal state, therefore, the conditions which must exist for a fibrinogen intoxication do not prevail.

I have above used the term 'poison the blood'; it will be advantageous for me to explain this expression.

The admixture of fibrinogen and blood may obviously affect the latter by causing it to clot or by preventing its clotting (*vide* previous papers),[1] but it produces other changes than these which are not so directly perceptible. The nature of these changes will be seen from the following: If in a normal dog the femoral vein be ligatured there is no obvious effect produced— *i.e.* there is no œdema of the leg. If, however, some solution of

---

[1] 'Intravascular Clotting,' *Roy. Soc. Proc.* 1886; 'Beiträge zur Frage der Gerinnung,' *Du Bois-Reymond's Archiv*, 1888; 'Ueber Schutzimpfung auf chemischem Wege,' *Du Bois-Reymond's Archiv*, 1888. [Collected Papers, pp. 135, 253, 329.]

PLATE 3.

To face page 352.

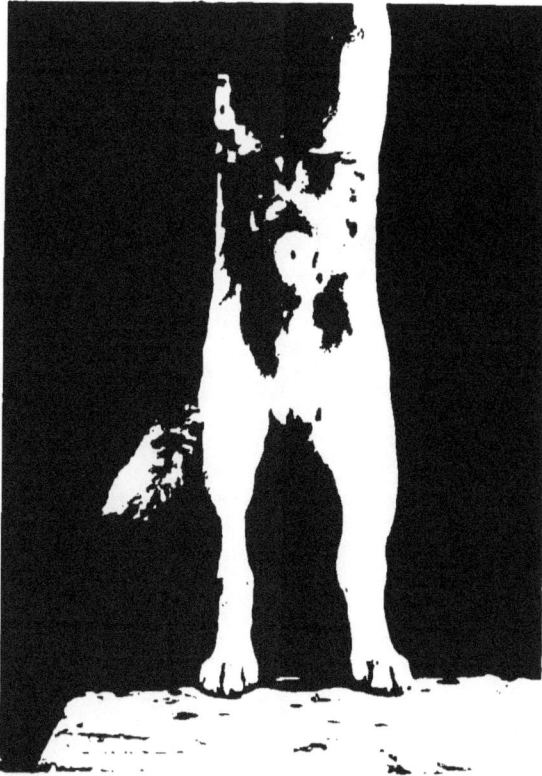

DESCRIPTION.

Photograph of a dog in which the left femoral vein had been tied at Poupart's ligament and tissue-fibrinogen solution injected into the left jugular vein. The oedema of the left lower limb is particularly noticeable in the leg about and above the ankle-joint.—V. H.

fibrinogen be injected into the circulation through the jugular vein, and the femoral be then ligatured, the effect produced is most pronounced, and is as follows : either the most extensive and rapidly developing simple œdema of the leg occurs or an enormous hæmorrhage '*per diapedesin*' takes place throughout the tissues of the limb; or the two are combined—there is hæmorrhage and œdema.

The injection of fibrinogen,[1] then, in addition to the obvious effects of clotting or delay in clotting, produces a totally disturbed relationship between the blood and the vascular wall, since, after the injection, a slight mechanical disturbance to the circulation causes a greatly increased exudation of the fluid .of the blood, or this associated with a free passage of the red corpuscles. The tendency the injection has to cause hæmorrhage I have already pointed out in a previous publication ; [2] the fact that it produces a simple but severe and sudden œdema is new. Now, to produce this altered state of the blood, leading to œdema, the same conditions of admixture of blood and fibrinogen are necessary—*i.e.* the admixture must be rapid. I will illustrate this by an experiment.

Exp. 3.—Used the NaCl fluid of thymus free from cells.

Dog I.—Weight, 17 lbs. 12 c.c. of solution rapidly injected into the jugular. Right femoral vein tied close to Poupart ligament. Dog killed the next day. The portal system thrombosed. The whole right leg extremely œdematous. Large hæmorrhages over upper part of leg and lower part of abdomen.

Dog II.—17 lbs. 12 c.c. of solution injected, but ten times diluted, and injection lasting five minutes. Femoral vein tied close to ligament. Dog killed next day. No trace whatever of clotting anywhere. Leg absolutely free from the slightest trace of œdema or hæmorrhage.

So far as my observations go, the tendency to œdema is the first symptom of fibrinogen intoxication—*i.e.* it is more easily produced than any other.

---

[1] The fibrinogen used to produce this effect may be lymph-fibrinogen, tissue-fibrinogen, or certain varieties of blood-fibrinogen.

[2] 'On Hæmorrhagic Infarction of the Liver,' *Trans. Path. Soc.* 1888. [Collected Papers, p. 346.]

A A

One of the most important features in these observations lies in their relationship to many important diseases. I have pointed out the conditions which must prevail to produce a fibrinogen intoxication. It is improbable that diseased conditions are often set up by a sudden large flow of lymph into the blood, but it is certain that the other condition, the slowing of the circulation in the neighbourhood of the thoracic duct, is a common incident; particularly I may mention valvular disease of the heart and obstruction to the circulation through the lungs, as conditions which necessarily produce this result. It is a dogma of medicine that cardiac dropsy, as a symptom of cardiac failure, is due to the mechanical obstruction of the circulation. My observations lead me to the conclusion that the danger in cardiac disease is fibrinogen intoxication; and that the symptoms of cardiac disease—e.g. dropsy, formation of intravascular clots, hæmorrhagic infarction, fever, &c.—are largely dependent on this condition.

PRINTED BY
SPOTTISWOODE AND CO., NEW-STREET SQUARE
LONDON